NORTHUMBERLAND
Place-Names

Published by Sigma Leisure - an imprint of Sigma Press, Stobart House, Pontyclerc, Penybanc Road, Ammanford, Carmarthenshire SA18 3HP.

British Library Cataloguing in Publication Data
A CIP record for this book is available from the British Library.

ISBN: 978-1-85058-994-5

Typesetting and Design by: Sigma Press, Ammanford, Carmarthenshire

Cover photographs: © Anthony Poulton-Smith
main picture: Prudhoe Castle
top (left to right): Haltwhistle sign; Angel of the North, Gateshead; Newbiggin-by-the-Sea sign; Duke of Northumberland public house sign, Newcastle-upon-Tyne

Photographs: © Anthony Poulton-Smith

Printed by: TJ International Ltd, Padstow, Cornwall

NORTHUMBERLAND
Place-Names

Anthony Poulton-Smith

SIGMA Leisure

Contents

Introduction

For years the history of England was based on the Roman occupation. In recent years we have come to realise the influence of the Empire did not completely rewrite British history, indeed there was already a thriving culture in England well before the birth of Christ. When the Romans left our shores in the fifth century the arrival of the Anglo-Saxons was thought to herald a time of turmoil, yet they brought the culture and language which forms the basis of modern England. Later the arrival of the Norsemen saw their influence and the same is true of our place names, the vast majority of settlement names in Northumberland are derived from the Saxon/Old English or Old Scandinavian tongues. There are also the topographical features such as rivers and hills still have names given to them by the Celts of the pre-Roman era.

Ostensibly place names are simply descriptions of the location, or of the uses and the people who lived there. In the pages that follow an examination of the origins and meanings of the names in Northumberland will reveal all. Not only will we see Saxon and Scandinavian settlements, but Celtic rivers and hills, Roman roads and even Norman French landlords who have all contributed to the evolution to some degree to the names we are otherwise so familiar with.

Not only are the basic names discussed but also districts, hills, streams, fields, roads, lanes, streets and public houses. Road and street names are normally of more recent derivation, named after those who played a significant role in the development of a town or revealing what existed in the village before the developers moved in. The benefactors who provided housing and employment in the eighteenth and nineteenth centuries are often forgotten, yet their names live on in the name found on the sign at the end of the street and often have a story to tell. Pub names are almost a language of their own. Again they are not named arbitrarily but are based on the history of the place and can open a new window on the history of our towns and villages.

Defining place names of all varieties can give an insight into history which would otherwise be ignored or even lost. In the ensuing pages we shall examine 2,000 plus years of Northumberland history. While driving around this area the author was delighted by the unique place names found in the county and so, having already taken a look at, among others, *Cambridgeshire Place Names* and *County Durham Place Names* turned here to Northumberland. This book is the result of the author's long interest in place names which has developed over many years and is the latest in a series which continues to intrigue and surprise.

To all who helped in my research, from the librarians who produced the written word to those who pointed a lost traveller in the right direction, a big thank you.

Place-Names

(in alphabetic order)

A

Abberwick

A name documented as Alburwic in 1170, this features a Saxon personal name and Old English *wic* and describes 'the specialised farm of a woman called Aluburh or Alhnurh'. In almost every case the speciality of a *wic* would have been dairy produce.

Acklington

First seen in 1177 as Eclinton, here is 'the farmstead associated with a man called Eadlac' where the personal name is followed by Old English *ing tun*.

Brainshaugh is a local name derived from 'the clearing of a man called Bregn or Bregwine'.

The locals enjoy a drink in the Railway, a pub name which can only have existed with the coming of the new mode of transport in the middle of the nineteenth century.

Acomb

Records of this name begin with Akum in 1268. This name is derived from Old English *ac* and describes the '(place at) the oak trees'.

St John Lee is recorded as Sancti Johannis de Leye in 1310, the name a permanent reminder of 'the church dedicated to St John in the woodland clearing'. This refers to St John of Beverley who had a hermitage nearby. He was Bishop of Hexham and, later, Bishop of York. Traditionally he was born to noble parents in Harpham, Yorkshire and educated at Canterbury and was a friend of the Venerable Bede, whom John ordained as a deacon and as a priest. It is said he performed many miracles, although little was recorded until his canonization in the eleventh century and later claims to his intervention, including that of Henry V who maintained it was John of Beverley who aided the English cause in the victory at the Battle of Agincourt.

Grottington is recorded as Grotinton in 1298, Grotyngton in 1479, Groteington in 1663, and Groatington in 1676, a Saxon personal name and Old English *inga tun* and telling of 'the farmstead of the family or followers of a man called Grotta'. The *wic* suffix in Keepwick is used in the sense of 'dwelling' and thus gives 'the dwelling of a man called Kepe'. Portgate is seen as Partyate in 1278 and as Portchet in 1382, this being from Old English *port geat* 'the way to the town'.

In early 2011 the newly refurbished Queen's Arms hit the headlines when the new pub sign was unveiled. The name is easily seen as supportive of the monarchy, a reference to the heraldic badge representing the office rather than a specific queen. Sign painters always portray the name as they see it. Together these two factors produced a sign which many found offensive.

When unveiled it appeared to be a caricature of the present queen. Smiling, her bare arms revealed a tattoo, one where the name 'PHIL' could be seen. The owner claimed it was based on his own mother, although he did concede there were similarlities between the two ladies. The owner's explanation does not explain the name on the tattoo, nor why the woman is standing in front of, what appears to be, a sketch of Buckingham Palace. Certainly no controversy where the Miners Arms is concerned, this area has a long association with mining. The Sun Inn offers a warm welcome, although an heraldic origin cannot be ruled out entirely.

Adderstone

Listed as Edredeston in 1233, this is derived from a Saxon personal name and Old English *tun* telling of 'the farmstead of a man called Eadred'.

Cocklaw has two possible origins for the first element, either *cocc* as in 'male birds' or the personal name Cocca, although the suffix is undoubtedly *hlaw* here meaning 'hill'.

Aggars Hill

Also seen in Aggarshill Fell, the earliest surviving document comes from 1278 where the name appears as Algerseles. This name is understood as featuring the Middle English element *schele*, thus 'the huts of a man called Alger'. This Saxon personal name is a pet form of Ealdgar or Aelfgar.

Akeld
Found as Echelda in a document from 1169, this is from Old English *ac helde* and describes 'the slope of the oak trees'.

Allen, River
First documented in 1275 as Alwent, this river has a British or Celtic name. Most likely derived from *albento*, describing the river as 'shining white'. The Allen is a tributary of the South Tyne.

Allendale Town
The earliest surviving record of this name dates from 1245 as Aleweinton. A name meaning 'the settlement in the valley of the River Allen', it is derived from Old Scandinavian *dalr* and Old English *tun* with a river name discussed under its own entry. The local pub also takes this place name.

Housty has two potential and quite similar meanings, most likely 'the sty for hogs' from *hogges stigu*, however *stigu* can also be used in a more general sense and thus this could be 'wooden enclosure for hogs'. Keenleyside is derived from 'the hillside by the woodland clearing of a man called Cenu', a Saxon personal name. Early records of Sipton Shiel show this to be 'the sheep stone shieling or summer pasture', although the meaning of the 'sheep stone' has never been understood.

Allendale's proximity to the Scottish border made this a troubled place to live during the sixteenth century. This resulted in fortification of farmhouses, these becoming known as 'bastles' – a name said to be derived from the French *bastille*. Not that anyone would ever recognise the famous Parisian stronghold in the bastles or bastle houses. Classically these would have very thick walls, around three feet was typical, with the bottom floor given over to stable space for the most valuable animals. Crucially there was no internal access between upper residential and lower stabling, windows were very small or mere slits, and the ladder pulled up when the family were safely ensconced in their home. Such places were remarkably secure.

Allenheads
Similar to the previous name this is the '(place by) the source of the River Allen'. Here the Celtic, or perhaps even pre-Celtic, river name

(discussed in detail under its own entry) is suffixed by Old English *heafod*.

This village stands near the southernmost tip of Northumberland. The modern village was founded on lead mining. One hydraulic engine built by William Armstrong, once used in the local sawmill, now stands in a purpose-built exhibition hall in the centre of the village. It is the only remaining engine of its kind which survives, several had been used in the lead mines.

Aln, River

A name recorded as early as the second century AD by Ptolemy and by the Venerable Bede in 720, documented as Alauvan and Alne respectively. Related to the French river names Allain, Aline and Allaine, this is from a Celtic adjective describing 'the bright or clear waters'. Whilst the Celtic term is unrecorded, it can be deduced by association with Gaelic *aluin* or *ailne* meaning 'fair, handsome' and to Welsh *alain* or *alwyn* meaning 'bright, clear, lucid'. Interestingly the river is found in a document dated 1539, where it is referred to as Water of Ale. Earlier the Venerable Bede refers to the meeting of the Synod of Twyford in 684, when the meeting was held at the confluence of the Aln and a small tributary which he names as Adtwifyrdi in his *Historia ecclesiastica gentis Anglorum*, although the modern location of Adtwifyrdi is unknown.

Alnham

Listed as Alneham in a record from 1228, this tells of 'the homestead on the River Aln'. Here Old English *ham* follows a Celtic river name discussed under its own entry.

Elilaw is a hamlet telling of 'the *hlaw* or hill of a man called Illa'. Scrainwood is difficult to define, although the first element certainly a personal name and possibly a nickname meaning 'chatterer'.

Alnham stands near the source of the River Aln and boasts a fourteenth century pele tower, restored in the nineteenth century. These pele or peel towers were fortified towers with an iron basket on top to light a signal fire – smoke during the day and fire at night – to warn of approaching trouble from across the border.

Alnmouth

Again the river mentioned here is the Aln, this name clearly telling this settlement was 'at the mouth of the River Aln'. Here the Celtic river name, discussed under its own entry, is followed by Old English *mutha.*

Pub names include the Hope and Anchor, a religious name taken directly from Hebrews 6:19 "Hope we have as an anchor of the soul". Naming the Schooner does refer to the coastal location and the two-masted sailing vessel. However it is also the name of a glass, one from which the locals would expect to drink sherry but when travelling to the USA and Australia this glass will contain beer. This same double meaning is also found in the Sun Inn, almost certainly of heraldic origin while also suggesting a warm welcome. As with the vast majority of coloured animal names, the Red Lion is heraldic. As the most common pub name in England it clearly refers to something or someone of great importance. The name increased in popularity following the union with Scotland, which it represented. It came here from an Englishman, John of Gaunt, whose bedroom antics resulted in an illegitimate line who went on to become rulers of Scotland and eventually England. While John of Gaunt never became ruler in his own right, he was undoubtedly the most powerful man in England during the fourteenth century.

Once Alnmouth was an important trading port, known for exporting great quantities of grain. A particularly violent storm in 1806 not only sounded the end for the already-deteriorating church on Church Hill, but changed the course of the river and ended the days as a port. The granaries which had held the grain awaiting the boats were no longer required and were converted into houses.

Alnwick

Another name pointing to the River Aln, a name discussed under its own entry. Here Old English *wic* is the suffix for the river giving 'the specialised farm on the River Aln'. The name appears as Alnewich in 1178.

Street names of this important old town show how it has grown over the years. One group recall some of the tradesmen of yesteryear, a theme which names those who would have worked in every village and one surprisingly rare in times when we see housing estates named after birds, trees and flowers in just about every town and city. Here the imaginative people have produced Weavers Way, Fullers Walk, Farriers Court, Tanners Court, Glovers Green, Coopers Close and Merchants Gardens. Pottergate also falls into this category, here the pots were likely sold although manufacture is also a possibility.

Bondgate Within and Bondgate Without show this *gata* or 'road' was lined by plots of land held by *bondi* or 'bondsmen'. Clive Gardens remembers Major-General Robert Clive, 1st Baron Clive and better known as Clive of India. Fenkle Street is a name found in many northern towns and comes from a word describing the 'bent' route followed by the road. The family of Sir Henry Percy, most often referred to as Harry Hotspur, have left their mark on the landscape in the form of Hotspur Hill, Hotspur Place, Hotspur Street, Percy Street, and Percy Terrace.

Cawledge Park puts together a Saxon personal name and Middle English *lache*, telling of 'the long swamp bordering the trickling stream associated with a man called Caua'. Deanmoor is a name speaking of 'the valley by the moor', from *denu mor*. Filbert Haugh is recorded since the thirteenth century, this being 'the *haugh* or enclosure of a man called Hildeburh'.

Hawkhill speaks for itself, however it is difficult to see if this represents 'the hill where hawks are seen' or whether this first element refers to a personal name. Hinding Burn and Hinding Flat share, with additions for the 'stream' and the 'flatlands', a name describing 'the valley frequented by hen birds'. Hulne comes from Old English *holegn*, a name recorded as Holen in 1296 and as Holne in 1334, 'the place where holly grows'. Plundenburn describes 'the valley stream where plums grow', the earliest known record dates from 1220 as Plundenburne. Shield Dykes means exactly what it says, recorded as Swynleys in 1288 this is 'the shilings or summer pastures by the dyke'.

Snipe House stands at a place recorded as Swinleysnepe in 1290 and as the modern form in 1663. It is fairly clear Snipe is a corruption, the name Sneap appears in several places nearby including a horseshoe bend on the Derwent. Among the Scandinavian languages of the Middle Ages we find two words, *sneap* is used as a rebuke or meaning 'to pinch', while *snopa* meant 'to castrate'. Here the origin must be to a sharp bend in a nearby stream, while we can assume the original place name was Swinley, itself telling of 'the woodland clearing where swine were reared'.

Pub names begin with Blink Bonny, named after a filly which hit the headlines in 1857 for both the right and the wrong reasons. As a two-year-old this blazed-faced and undersized racing horse won eight of its eleven starts. Next year it started a further twenty races, winning fourteen. Among its wins were the Derby by a neck and the Oaks by an impressive eight lengths. Jockey John Charlton was blamed for her defeat in the 1,000 Guineas and the St Leger, indeed it was claimed he had been paid to throw the race especially as Imperieuse was the victor on both occasions.

The Plough Inn is amongst the most common of pub names. Dating from the days when almost everyone earned their living off the land, it was a clear invitation to agricultural workers. Tanners Arms may have originally been named by a landlord who had previously worked in leather tanning. The Fleece Inn is a reminder of the importance of the wool trade in England. Following the enclosures and the introduction of vastly improved spinning and weaving techniques, the British economy was so strong it laid the foundations for centuries of colonial rule and empire. Others ignore the product and commemorate those who tended the flocks such as in the Shepherds Rest.

The George Inn is a patriotic name, one showing support for the monarchy in general by citing the name of four Hanoverian kings. The Queens Head is oldest pub in Alnwick, dating from the sixteenth century and showing a sign with the unmistakeable image of Elizabeth I. Another patriotic symbol hangs outside the John Bull Inn, this being the very epitome of Englishness created by John Arbuthnot in 1712.

For the origin of Grey's Inn we need look no further than the castle. Seat of the Percy family it was also the birthplace of Sir Thomas Grey, a long military line associated with the county. While the Bluebell is very often related to the woodland spring flower, the original name marked a connection to the local church. The Oddfellows Arms marks the meeting place of a benevolent society founded in the nineteenth century. Even earlier the Olde Cross would have been a waymarker at a junction or crossroads. Travellers would have been guided by the signpost and also by the sign hanging outside the Farriers Arms if they had had a horse to care for too. What was once the Black Swan was changed to the Hairy Lemon by new owners, said to have been inspired by a pub in Dublin but with no record of its etymology.

Alwin, River

Listed as Alewent in 1228, the name having identical origins to that of the Alne and describing its 'bright, clear waters'.

Alwinton

Listed as Alwenton in 1242, this comes from Old English *tun* and a Celtic river name discussed under its own entry and giving 'the farmstead on the River Alwin'. This name is also found as Allenton in some documents, this representing the pronunciation of the place as it is still known today.

Clennell describes itself as 'the clean hill', a reference to the lack or vegetation. From Old English *thyrne ham* comes Farnham or 'the homestead by the thorn bushes'. Foxton is a modern corruption, for the record of 1324 gives Foxden from *fox dene* 'the valley frequented by foxes'.

Howsdon Burn is a stream which takes its name from 'the hill of a man called Holl'. Allerhope Burn takes the name of the valley through which it flows as 'the valley growing with alder trees'. Netherton is a common place name, Old English *neother tun* 'the lower farmstead'. Ridlees comes from dialect *ridd* and Old English *laes* and tells of 'the pastures cleared or ridded'.

One local was doubtless inspired by its location. Less than five miles from the border, the Rose and Thistle represents England (in the rose) and Scotland (with the thistle). This location has also given a name to the Alwinton Border Shepherds Show, a fair for sheep farming held on the second Saturday of October. As the last agricultural show of the year in the area, it signals the end of summer and shepherds and farmers return to their flocks and prepare for winter.

Amble

Records of this name include Ambell in 1204 and as Anebell in 1256. A Saxon personal name is joined by Old English *bile* and speaks of 'the promontory of a man called Amma or Anna'.

Gilden Burn has been seen since the twelfth century and tells of 'the valley of a man called Gildwine'.

Close proximity to, and a long history associated with, the sea is seen in the names of two pubs: the Harbour Inn and the Dock. The Schooner Inn is a third example, although it also represents the name of the glass in which sherry is traditionally served. The Masons Arms is a reminder of one of the oldest guilds of craftsmen, later becoming a secret society who may have gathered at places so named. Note the Blue Bell Inn has nothing to do with the spring flower. Both the image on the sign and the name tell us this, the flower is the bluebell and but one word. Here the reference is the church, these premises likely standing on church land.

Coastal erosion had revealed a local seam of coal by the reign of Elizabeth I. For some time attempts were made to turn Amble into a port to ship locally produced coal and salt. All these failed, probably due to the success of larger mines in the region. Yet this did not stop William Hewitt from retaining the mining rights when he sold his holdings in the village in the middle of the seventeenth century.

Ancroft

The earliest surviving record of this name is as Anecroft in 1195. Here Old English *ana croft* tells of 'the isolated enclosure'.

Locally we find Allerdean, a name found since the thirteenth century and derived from 'the *dene* or valley of a man called Aelfhere'. The name of Edmondhills has never been defined with any certainty, however there are two likely meanings, neither of which features an Edmond or any other personal name. Thus this would seem to be either *eamot hyll* 'the hill where the rivers meet' or *amette hyll* 'the hill frequented by ants'. While 'the *tun* or farmstead of a man called Heardgar' features a personal name unrecorded in Saxon England, however we can be fairly confident this is the correct definition of Haggerston as it does appear in related Teutonic languages.

Population in Ancroft has fluctuated over the years. One period of growth in the seventeenth century coincided with the return of the plague. In their ignorance as to the source of the disease the victims were taken to a field south of the village, their last days spent in rudimentary huts made from the branches cut from the nearby broom trees. These then became their final resting place when the body and branches were burned together where they lay. To this day the field is known as the Broomies.

Anick

Found as Aeilnewic in 1160, as Einewic in 1225, Ainewik in 1226, Anewyke in 1296, as Aynewyk in 1479, and as Anyk in 1536, this name comes from Old English and describes 'the specialised farm of a man called Aethelwine'. Here the Saxon personal name is followed by *wic*, here given as 'specialised farm' which in almost every case refers to a dairy farm.

As astonishing a pub name as is found anywhere in the country is found here. This old drovers inn was certainly here by the early eighteenth century and was known by its present name. Its history is well documented although the name is a complete mystery.

Of course there are local explanations for the name. One suggests this was a meeting place for rat catchers, which hardly seems likely as how many rat catchers would a place the size of Anick require? Another tale tells this was where the largest rat ever found was caught, again this hardly seems plausible as surely no inn-keeper would want to advertise vermin no matter how large. The third narrative tells how a landlord gave information to the officers of the Crown during the Jacobite uprisings of 1715 and 1745, earning the name 'The Rat'. This is the least plausible of the three for it is clearly a case of creative etymology.

Ashington

Found as Essendun in 1205, this is from Old English *aesen denu* or 'the valley where the ash trees grow'.

Street names come from a number of sources. When coal mining brought workers to the town, housing quickly became a problem and the grid system still evident was born. The simplest names to explain are those running east to west. These avenues began with First Avenue nearest the town centre and ended with Seventh Avenue furthest away, with five other quite obvious names between. Nearby are streets with a theme of trees and include Beech Terrace, Redwood Court, Hawthorn Road, Chestnut Street, Sycamore Street, and Maple Street.

There are also those named after Shakespearian characters, all of whom are female. Best known would be the eponymous character who gave her name to Juliet Street, she the thirteen-year-old daughter of the House of Capulet in *Romeo and Juliet*. Portia Street takes the name of the rich heiress in the play *The Merchant of Venice*. Katherine Street takes the name of Kate Minola, the woman referred to in the title of *The Taming of the Shrew*. Beatrice Street remembers the leading lady and lover of Benedick in *Much Ado About Nothing*, their witty remarks providing much of the entertainment.

Haldane Street is named after Lord Haldane, former Liberal turned Labour politican, lawyer and philosopher who became Secretary of State for War from 1905 until 1912. He then served as Lord Chancellor until 1915 when he was forced to resign, when accused of having German sympathies, which he furiously denied and has never been proven. Bowman Square recalls after Sir James Bowman, who worked at Ashington colliery from the age of fifteen, returning for a second stint after serving his country in the Royal Marines during World War I. He rose to become General Secretary of the Miners Union by 1935, helped to organise the German trade unions in the aftermath of World War II, before taking the post of chairman of the National Coal Board in 1956 until 1961 when he was forced to retire due to ill health.

To the south we find the region known as Stakeford. Situated to the south of the River Wansbeck, it is reached from Ashington by the West Ford Road, today crossing the river via a bridge. The name of the place means 'the ford marked by stakes', however this is not a 'ford' in the modern sense but was a drier path through marshland. Designed to show the depth of the marsh, the stakes can still be found in places, surely one of the very few modern places names where the man-made origins are still visible.

One local pub here is the Rohan Kanhai, he appearing here when he was captain of the West Indies cricket team. Born Boxing Day 1935, Rohan Bholalall Kanhai was a batsman of great talent and consistency. His first three test matches, while touring England in 1957, saw him keep wicket and open the batting. However it was as a middle order batsman he is remembered, his 79 tests seeing him average 47.53 with a career best of 256. Not that he was ever seen as a traditional test match cricketer – indeed his patient and steady half century in the inaugural world cup final of 1975 was quite out of character, although it enabled Clive Lloyd's explosive hitting to build a match-winning total – some of his shots were most unorthodox, including the 'falling hook shot' which saw him end up on his back and the ball invariably hitting the boundary ropes. Kanhai also played county cricket for Warwickshire, alongside fellow West Indians Alvin Kallicharran, Derek Murray and Lance Gibbs and behind England openers Dennis Amiss and John Jameson his 11,615 runs at an average of 51.62 made him the highest run scorer for the county, a record which was to last until the record books were re-written by countryman Brian Lara.

The name of the Black Diamond Inn is not as obvious as it once was. Once this was a phrase comparable to a 'rough diamond' today, a person of good morals if lacking in etiquette. As both coal and diamonds are associated with carbon, the name is often seen as a reference to coal mining. Mining has also produced a distinctive accent and dialect, noticeably different to Geordie. It is officially known as Pitmatic, although the locals refer to this as 'yakka'.

The Fox Cover Tavern is a misnomer, one taken from an existing place name which was 'fox covert', a place where foxes were known to be found. Woodhorn Grange takes the name of the former colliery, itself named for a place speaking of 'the horn of land where woodland grows'. Until a victim of the Beeching axe in 1964, the local station was known as Hirst. This minor place name comes from *hyrst* 'wooded hill' and hence the origin of the Hyrste Castle public house.

Ayle

A name which takes the name of the Ayle Burn, the river on which it stands which is found as Alne in 1347. This is a Celtic river name and one which has never been understood.

B

Bamburgh

With the earliest record dating back to the early eighth century where it appears as Bebbanburge, this is derived from a Saxon personal name and Old English *burh* and referring to 'the stronghold of a queen called Bebbe'. The most obvious feature here, Bamburgh Castle, was considered the castle Joyous Gard, always associated with Sir Lancelot.

Bamburgh Castle had stood even before the arrival of the Germanic tribes. The earlier fortification, then known as *Din Guardi*, marked this as the capital of the kingdom of Bryneich. When it became another of the Anglo-Saxon kingdoms it was known as Bernicia, Bamburgh remaining its capital, until the unification with the neighbouring kingdom of Deira in 634. Henceforth the kingdom was known as Northumbria with its capital the city of York showing the kingdom was much bigger than the present county.

Blackmiddingmoor is as long a single name as found anywhere in England. It features three elements *blaec middyng mor* or 'the black midden swamp', a midden being where all waste was dumped, mostly organic simply because that was all they had except for a few pots, and it did include human waste. The Mizen Hotel probably shares the middle element with a slurred pronunciation of 'midden'. The most imposing building in Bamburgh also gave a name to the Castle public house.

Burton is one of the most common English place names, from *burh tun* it speaks of 'the fortified farmstead'. Fleetham comes from *fleot ham* and tells us of 'the homestead by the tidal stream'. Hoppen, found as Hopum in 255 and Hopune in 1296, is thought to represent *att hopum* meaning the '(place) at the enclosed valley'. Monkshouse is a lasting reminder of 'the storehouse of the monks' which once stood here.

Mousen, found as Mullesfen in 1166, speaks of this as 'the fenland associated with a man called Mule'. Here the Nanny River is first found in 1245 as Nauny, doubtless this is a Celtic river name and earlier records are needed before the name can be defined. Outchester can be traced back to 1236 when the name is recorded as Ulecestr, this referring to 'the Roman stronghold frequented by owls'.

Ratchwood is found as Wrethewode in 1279, however the listing from 1620 as Wretchwood is easier to see as 'the wood of the outlaws' from Old English *wreccean wudu.* Rosebrough sees the suffix used in a manorial sense, here this is 'the dwelling of a woman called Osburh'. One of the most common place names in the land, Bradford comes from *brad ford* Old English for 'the broad ford'. The clearest early example of Shoreston comes from 1191 as Shotesdon, itself from 'the *dun* or hill of a Scot'. Tughall is recorded as Tughala in 1104 and as Tuggahala in 1175, a name describing the '*halh* or corner of land of a man called Tugga'.

Barrasford

Listed as Barwisford in 1242, this is from Old English *bearu ford* and describes 'the ford by a grove'.

Bavington (Great & Little)

Two places with a common ancestry recorded as Babington in 1240. Here is a Saxon personal name and Old English *ing tun* and referring to 'the farmstead associated with a man called Babba'. The additions are self-explanatory.

In 1794 Bavington school had a new teacher for its single class. Later to earn fame in mathematics and astronomy, Henry Atkinson should probably be remembered for this position too. He was just 13 years of age at the time.

Beadnell

Listed as Bedehal in 1161, this comes from a Saxon personal name and Old English *halh.* Together these described 'the nook of land of a man called Beda'.

The Coach House Inn shows this was the local port of call for travellers along the many coaching routes across the country. Travellers from the sea needed to be aware of the nature of the coastline here, for Beadnell's port has the unique claim as the only port on the east coast with a harbour entrance facing west.

Beal
The earliest surviving record of this name dates from 1208 where it appears as Behil. Derived from Old English *beo hyll* this name describes 'the hill frequented by bees'.

Beanley
Seen in the middle of the twelfth century as Benelega, this comes from Old English *bean leah* and describes 'the woodland clearing where beans grow'.

Bedlington
Records of this name include Bedlingtun around 1050. This comes from Old English *ing tun* and a Saxon personal name and describes 'the farmstead associated with a man called Bedla and Betla'. Bedlington was once an exclave of County Durham, once a common if confusing feature throughout the country, and one brought about by land ownership. This led to the parish being known informally as Bedlingtonshire.

Street names begin with another link to the neighbouring county. Berrington Court takes the name of Shute Berrington who, after terms as Bishop of Llandaff and Bishop of Salisbury, was installed as Bishop of Durham in 1791. He ordered troops to Chester-le-Street in 1812, to break up a strike by miners, a reminder of the powers then still held by the Prince Bishops of Durham which entitled them to a private army.

Modern streets tend to follow themes, such as that of trees seen in the region of Thorn Tree Drive, Yew Tree Drive, Ash Tree Drive, Rowan Close, and Cedar Close. Yet some are mercifully more imaginative and the Romano-British culture from the first half of the first millennium AD has resulted in Centurion Way, this the Roman leader of eighty men. Note this is twenty less than the number normally quoted, this earlier belief based on the idea that 'centurion' came from *centum* the Latin for 'one hundred'. This is now widely disputed.

Other streets with a 'Roman' theme are Augustus Drive and Octavia Close. These named after the first Roman emperor whose full name was Imperator Gaius Julius Divi Filius Caesar Octavianus Augustus. Corchester Road takes the name of a Roman station on Hadrian's Wall, most often referred to as Coria or Corbridge. This Roman place name has been linked to others in the street names of Winchester Way and

Chichester Grove. All three feature the Old English suffix *caester* or 'Roman stronghold'. It was once thought this came from Latin *castra* and yet, although they are undoubtedly from the same root, not to have a Saxon word in a Saxon place name makes no sense. Furthermore, without *caester* being used in these place names, this Old English word will have existed but never used! There is another 'chester' road name here. Borchester Close would seem to be another Roman place name, yet there never has been a place on the UK map named Borchester – although it does exist. This is the fictional nearby town featured in the long-running radio series *The Archers*, a popular shopping place for the Ambridge residents.

Sheepwash speaks for itself as 'the place for washing sheep'. Some pub names took on minor place names too. Both the Ridge Farm and Bank Top can be seen as referring to features in the landscape, while the Market Tavern would have doubled as a meeting place for the organisers of the lucrative market.

While coloured animals are invariably heraldic, the Dun Cow probably refers to a well-loved local animal. However the Black Bull, an image taken by many families as a symbol of strength, would have been found on a coat of arms. The Red Lion is another which fits into this category, the most common pub name in England is most often representative of Scotland. Another animal gave its name to the Bedlington Terrier, this breed of dog being named after the village.

The Clayton Arms is named after the man who did much to preserve Roman artefacts and remains associated with Hadrian's Wall. On Burdon Terrace, named after the family who were lords of the manor for more than a generation, stands the Wharton Arms, itself after the Wharton family who also served as lords of the manor. Born at Ford Hall, Bishopwearmouth, Henry Havelock gave his name to the General Havelock. His military career was most notable for his time in India and in particular his role in the Indian Rebellion of 1857. The Percy family, earls of Northumberland, gave their name to the Percy Arms and to both the Northumberland Arms.

Belford

Listed as Beleford in 1249, Belleford in 1301, Beleford in 1313, Belford in 1314, Belforth in 1323, Belfurthe in 1460, Belforth in 1550, and Belford in 1610, all of which point to 'the ford and/or valley of a man called Beola or Bella'.

Locally we find Ross, a name of Celtic origin. This culture had no written form, which would make understanding impossible were it not for related tongues which have survived and do have written forms. Here Old Irish *ros* and Welsh *rhos* describe 'the high wasteland'.

There are a good number of pubs in England named the Black Swan. Today we know these birds are native to Australia, however the name is seen well before Captain Cook arrived in the antipodes in the late eighteenth century and thus cannot be attributed to this discovery. The phrase 'black swan' can be traced back at least to the Roman poet Juvenal, who used it to refer to a *rara avis* 'a rare bird' and thus describing a remarkable individual, one used as for a pub name by the sixteenth century and probably much earlier.

Nearby we see Spindlestone Ducket Mill. This tapering stone tower with a conical roof was built in the eighteenth century but rather oddly nobody seems to be able to agree to its original use. Some maintain it was built as a windmill tower, while others claim it was a dovecote.

Bellingham

The earliest known documented record of this name comes from 1254 where it appears exactly as it does today. Where names have not changed it is difficult to see the meaning and here there are two possible meanings. Either this is 'the homestead of those living at the bell shaped hill' from Old English *belle inga*, or this could be understood as 'the homestead at the bell shaped hill'.

Garden House takes its name from Old French *gardin*, first seen in

above: Bellingham's communal centre *right*: Gingall Gun from the NW Fort at Taku, China presented to the people of Bellingham

1279 as Gardina and referring to an ornate garden. Hesleyside tells us of 'the hill where hazel grows'. The churchyard of St Cuthbert's has a grave known as the Long Pack. Locally held to be the last resting place of a burglar, one who had hidden in a beggar's pack in order to gain access to the household. He was discovered when racked by a fit of coughing, the householder running him through with his sword.

Belsay

A name derived from Old English *hoh* with a Saxon personal name, this describes 'the hill spur of a man called Bill'. This name is recorded as Bilesho in 1163.

Belsay Castle's first wave of construction was completed around 1370 when it was home to the Middleton family. The continued to live here until it was abandoned in the early nineteenth century in favour of the newly-built Belsay. Much of the interior was taken away, no doubt some reappeared in their new home, leaving the castle to act as the family's folly, which no self-respecting aristocrat would be without at the time.

Berwick-on-Tweed

From Old English *bere wic*, this basic name means 'the barley farm'. The addition comes from the local river, the name is discussed under its own entry. This place is recorded as Berewich in 1167 and as Berewicum super Twedam in 1229.

Public houses originally displayed a sign which, like every other retail establishment, advertised the product when the majority of the population were illiterate. Few names referring to the product are more common than that of the Brewers Arms. The product was the original reason for the pub sign and continues to be advertised in such signs as the Barrels and also the Barrels Alehouse.

The Leaping Salmon is named for the Tweed, one of the most important and productive salmon rivers in the land and also the basis for the Fishers Arms. The White Swan Inn most likely refers to the majestic bird, although it is a popular heraldic device too. A coastal location has given us the Pilot Inn, today associated with air travel but which began as the man who used his local knowledge to bring ships into a harbour. Similarly we find the name of the Ship Inn.

The White Horse is a popular heraldic image, but as a pub name most often represents the House of Hanover, this royal family ruled from George I in 1714 until Queen Victoria in 1901 – 187 years and just six monarchs. The latter is the basis for the Queens Head and also the Victoria.

The Salutation Inn is a religious name, this referring to the Annunciation and a celebration of the visit by the Archangel Gabriel to the Virgin Mary to announce she had been chosen to be the mother of the Son of God. Location was also the reason for the name of the Elizabethan, it taking the name of the Elizabethan town walls. The Cross Inn stands on the site of an old marker, the Meadow House is self-explanatory. Until banned in the middle of the nineteenth century, pubs were often the venue for bear baiting. This is the most likely reason behind the Brown Bear.

Trades are a popular theme for it acts as an advertisement for those whose work is associated with these people. Among the earliest crafts and guilds is that seen in the Masons Arms, while the Plough invites those working the land which, when the name was first popular in the sixteenth century, meant virtually everyone.

History is an excellent potential source for pub names. In 1704 a battle of the War of Spanish Succession, the British forces led by the future Duke of Marlborough, was fought at a place which gave its name to his seat of Blenheim Palace and locally to the Blenheim public house. In the 1840s the Liberal Party adopted a policy of international trade free from restrictions such as quotas and import duties. These were put in place in 1860 and Free Trade became a pub name even though this legislation was reversed in 1932.

Berwick's history as sometimes an English town and other times in Scotland, led to an oddity in the passing of parliamentary legislation. Despite the Act of Union of 1707, England and Scotland have quite separate legal systems. In order to clarify that the English system also covered Wales, an Act was passed in 1746 deeming such and, as the title of the Wales and Berwick Act 1946 implies, also covered Berwick-on-Tweed. Note Berwick-on-Tweed remained a county in its own right until 1885, not included in Northumberland until 1885 and then solely for parliamentary purposes.

Biddlestone

Recorded as Bidlisden in 1242, this comes from Old English *bythle denu* and refers to 'the valley with a building'.

Bilton

Early records of this name include Bylton in 1242. Here a Saxon personal name and Old English *tun* speaks of 'the farmstead of a man called Bill or Billa'.

Bingfield

This place name is from Old English, either a Saxon personal name with *inga feld* and 'the open land of a man called Bynna' or if the first element *bing* then this would be 'the open land with a hollow'. This name is recorded as Bingefeld in 1181.

Grindstone Law speaks of the 'hill where grinding stones are quarried'. Shelbraes is a corruption of the seventeenth century Schellawe, from *shiel hlaw* meaning 'the hill with summer pastures'. Todridge is derived from Old English *tod hrycg* and, recorded as Todrige in 1479, tells of 'the ridge of land frequented by foxes'.

Birling

Listed as Berlinga in 1187, this comes from a Saxon personal name and Old English *inga* and describes the '(place of) the family or followers of a man called Baerla'.

Birtley

Recorded as Birtleye in 1229, here is a name from Old English *beorht leah* and describes 'the bright woodland clearing'.

The spelling of Buteland has been largely unchanged since the thirteenth century when it described 'the cultivated land of a man called Bota'. Listed as Dunley in 1279, the name of Dinley comes from *dun leah* 'the woodland clearing by the hill'.

Enjoying a drink at the Percy Arms probably does not mean the customer immediately thinks of the powerful family who were dukes of Northumberland from 1377.

Blanchland

Seen in a document from 1165 as Blanchelande, this is one of the handful of place names derived entirely from the Old French language. This name speaks of 'the white woodland clearing' from *blanche launde*.

Acton is a hamlet telling of 'the oak tree farmstead' from Old English *ac tun*, the earliest record is seen in a document dated 1255 as Aketon. Beldon Burn is a name from the stream here, derived from a combination of Old French *bel* and Old English *denu* 'the beautiful valley'. Newbiggin is an Old English name for 'the new building' which also tells us it was not the original and thus there was a settlement here well before the first reference of Newbigging in 1378. Sessinghope is found as Sessynghop in 1336, Sessinghope in 1364, Cissyhope in 1538, and Cessinghope in 1595. This is from Old English *inga hop* and a personal name and telling of 'the valley of the family or followers of a man called Cissa'.

Named after a former Bishop of Durham, the Lord Crewe Hotel dates from around 1165. With hidden stairways and rooms, this has been a hiding place for monks for much of its life. One of the buildings best features is its huge fireplace. During the 1715 Jacobite uprising one General Tom Forster hid in this fireplace to evade capture. His sister Dorothy is reputed to haunt the place.

Blyth

A name coming from the local river, a common river name found in several places in the country. This comes from Old English *blithe* and describes 'the gentle one'. As a place name it is recorded as Blida in 1130.

Street names here include those remembering the politicians who have contributed to the history of our islands. Gladstone Street recalls William Ewart Gladstone, a politician whose sixty-year career included four terms as prime minister, he was the oldest ever to hold this post when he finally resigned, and four as chancellor of the exchequer. Disraeli Street recalls Gladstone's great rival, Benjamin Disraeli, 1st Earl of Beaconsfield. His two terms as prime minister shaped the modern Conservative party,

Salisbury Street recalls the Marquis of Salisbury, three of whom served as prime minister although this refers to the third holder of the title who served three terms as prime minister in the late nineteenth century. Balfour Street comes from Arthur James Balfour, another who served as prime minister and as foreign secretary. He is best remembered for his role in the political situation in Ireland and the Entente Cordiale agreements with France.

Portland Street takes the name of William Cavendish-Bentinck, 3rd Duke of Portland, whose two terms as prime minister in the late eighteenth and early nineteenth centuries saw a gap of 24 years between them, the longest between terms of office of any prime minister in this country. Goschen Street remembers George Goschen, 1st Viscount Goschen, whose twenty-five years in politics saw him serve as chancellor of the exchequer for over five years. Keelman's Terrace doubtless refers to those who worked on the boats hauling coal from the coast to waiting boats which could not use the rivers themselves owing to the long keels of these vessels which would have grounded them in the too-shallow rivers.

Bridge Street leads to Waterloo Bridge, the latter named as it was built in 1814, just two years after the famous battle. Without doubt the leading family in the Blyth area in recent centuries are the Ridleys. Their name can be found in Ridley Street and Ridley Place, with Blagdon Street named after their home. Turner Street took the name of the landholder, whose contribution to the port's history is better remembered as a ship owner. Gray Street honours Lord Gray, a draper of Blyth who later entered politics and served as mayor of Hartlepool. Other local businessmen, all of them solicitors, inspired the names of Sydney Street, Lynn Street, and Marlow Street.

Horton is a common name and one always referring to 'the muddy farmstead' from *horh tun*.

Locals enjoy a drink in the Flying Horse. As a pub name it is invariably depicted as the winged horse Pegasus from Greek mythology, yet the true reference is an heraldic pointer to the Knights Templar. At the Prince of Wales the sign shows an image which is held to be that of the future Edward VII, the image of the Prince of Wales feathers is etched into the glass for the windows.

The Oddfellows Arms shows this was a meeting place for the Independent Order of Oddfellows, a benevolent society founded in the early nineteenth century with branches in the UK and many other countries. Three Horseshoes seems an unusual name for a pub for, as even the youngest child soon discovers, horses have four legs and thus require four shoes. However it makes more sense if seen as a question, then the answer can be found at the inn where the nearby blacksmith took care of the horse while the rider rested and refuelled inside the inn.

The Royal Tavern is a general reference to the monarchy and shows this to be a patriotic establishment, as does the Kings Head. Patriotism was the reason for naming Waterloo Road, the famous battle of 1815 to end the Napoleonic troubles also giving a name to the Waterloo which stands on a road leading off its namesake. Similarly the Isabella is near the junction of Isabella Road. The Bebside is named after a minor place name, as is the Newsham.

The Blyth and Tyne, a pub with its name picked in the brown and cream tiled frontage, takes the names of the two most important rivers to the area. Such a waterside location is suggested in both the Boathouse Tavern and the Steamboat Inn. Employing an expression which has come to us from across the Atlantic, the Time Out offers a welcome break for refreshment.

Chicago businessman George Pullman owned the Pullman Carriage Works, an engineering firm who raised the standard in the production of railway coaches. The Pullman is one of several reminders of the railway here. Certainly one of the skills needed to produce a quality coach would have been those of the craftsmen who gave a name to the Joiners Arms. Another of the earliest known craftsmen gave a name to the Masons Arms.

The Black Diamond is a reference to coal, mining being an important industry in this area for many years. No doubting the Windmill Inn points to this being the location of that early technology.

From the story told in Greek mythology, the Golden Fleece recalls the Argonauts who battled many trials and tribulations to recover this prize. This image was present in the arms of the Knights of the Golden Fleece, an order of chivalry founded in 1429 to protect the church. Nothing Greek about the Spartan, however. Here the reference is to the local football team, Blyth Spartans a non-league team best-known nationally for their exploits against league teams in the F A Cup.

The Seahorse may well be heraldic, while its proximity to the sea will have been a factor this image appears in innumerable coats of arms. Another likely heraldic image is the probably origin for the Willow Tree, seen in many arms including that of a fourteenth century Bishop of Durham.

Dukes of Northumberland, the Percy family are amongst the most powerful families in English history, hence the name of the Percy Arms. A family related to the dukedom gave a name to the Sidney Arms, Sir Philip Sidney being the grandson of the John Dudley, 1st Duke of Northumberland through his mother. The Golden Eagle is another popular heraldic device, such a strong image has been used by innumerable families and some of the most powerful nations on the planet, such as Russia, the USA, and the Roman Empire.

Probably the best-known landmark in Blyth is the well-named High

Light Lighthouse. Began in 1788 it was initially just 35 feet high. Exactly a century later another 14 feet was added with a final 12 feet in 1900. At over sixty feet in height the lighthouse ended its working life in 1985 and was given Grade II listing two years later.

Bolam

This name is found as Bolum in 1155, a name derived from a plural form of either Old English *bola* or Old Scandinavian *bolr*, both of which describe the '(place at) the tree trunks'.

Harnham is a local name from *hyrne ham* or 'the homestead in a corner of land'.

Bothal

Listed as Bothala in the twelfth century, this name speaks of 'the nook of land of a man called Bota'. Here the Saxon personal name is followed by Old English *halh*.

Longhirst is from Old English *lang hyrst*, quite literally 'the long wooded hill' and describing a ridge of land covered by trees. Old Moor and New Moor are self-explanatory, other than the respective ages refer to when these lands were first utilised.

Boulmer

Found as Bulerner in 1161, here Old English *bula mere* speaks of 'the pond frequented by bulls'.

During the eighteenth and nineteenth centuries Boulmer was notorious for its smugglers. One character, the infamous William Faa, king of the gypsy community, hailed from the village of Kirk Yetholm in Scotland. Boulmer does not have a harbour, the fishing boats called cobles are hauled directly on to the shore to land their catch of crabs, lobsters and sea salmon.

Bowsden

A record from 1195 as Bolesdon shows this to be from a Saxon personal name and Old English *dun* and 'the hill of a man called Boll'.

Brandon

Records of this name include Bremdona in the middle of the twelfth century and as Bromdun in 1236. Here is 'the hill where broom grows' from Old English *brom dun*.

Branton

Derived from Old English *bremen tun* and speaking of 'the farmstead overgrown with broom', this name is recorded as Bremetona in the middle of the twelfth century.

Branxton

Here a Celtic personal name and Old English *tun* combine to tell us of 'the farmstead of a man called Branoc'. The earliest record of this name dates from 1195, appearing as Brankeston.

Nearby Piper Hill has a granite cross commemorating the site of the Battle of Flodden on 9th September, 1513. The invading Scots under King James IV were defeated by English forces led by the Earl of Surrey in what is still sometimes referred to as the Battle of Branxton.

Breamish, River

Listed as Bromic in 1050, as Bremish in 1532, and as Bromish in 610, this is a British river name related to Welsh *brefu* meaning 'to roar'. Clearly this is a description of the young river, showing this was where the river was named.

Brinkburn

A name listed as Brinkeburne in 1120, Brincaburch in 1175, Brenkeburn in 1313, Brenkborne in 1542, and Brenckburn in 1728. A varied list of forms which show the place is named from the river. However the forms do not show if this is 'the stream below the brink (hill)' or 'the stream of a man called Brynca'.

Minor names include Oakhaugh, *acen haugh* describing 'the mound growing with oak trees'. Lynch Wood comes from Old English *hlinc wudu* 'the wood on the ridge of land'.

Brunton

Found as Burneton in 1242, here is 'the farmstead by a stream' from Old English *burna tun*.

Buckton

This place name is first found in the early thirteenth century where it appears as Buketun. This record is later than we would hope, meaning we are unable to distinguish between the three possible origins. If this is Old English *bucc tun* it speaks of 'the farmstead where male deer are kept', if *bucca tun* this would be 'the farmstead where male goats are kept', or if the first element is a personal name this would be 'the farmstead of a man called Bucca'.

Budle

From Old English *bothl* and describing 'the special building', this name appears as Bolda in 1166.

Burradon

Found as Burhedon in the early thirteenth century, this is derived from Old English *burh dun* and tells of 'the hill with a fortification'.

Buston (High & Low)

Records of this name include as Buttesdune in 1166 and as Butlesdon in 1249. There are two possible origins here, either this is a Saxon personal name and Old English *dun* and tells us it was 'the hill of a man called Buttel'; however if the first element here is *butt* then this would be 'the stumpy hill'.

Bywell

Listed as Biguell in 1104 and as Biewell in 1195, this comes from Old English *byge wella* and describes 'the spring near the bend in the river'.

Apperley is easy to see as 'the clearing by the apple trees', from *aeppel leah*. Nearby Backworth describes itself as 'the *worth* or enclosure of a man called Bacca'. From *brom halh* comes Broomhaugh, 'the nook of land where broom grows'. Broomley is similar and features the suffix *leah* giving 'woodland clearing where broom grows'. Fortherley is first seen in 1255 as Falderleg, a name from Old English *falder leah* or 'the woodland clearing with a fold for sheep'.

Healey *hean leage* or 'the high clearing', while Hindley represents 'the woodland clearing frequented by hinds'. Early forms of Kipperlynn are as Skitterlyn, this seems to be from *skitter* literally 'to void excrement' and understood to be a name of contempt for a very muddy or dirty place. Most of the early forms of Minsteracres are quite similar in appearing to describe land belonging to a large church, yet there is no evidence to support this. However there are records, such as that of Mynstanesacres in 1268, suggesting this is 'the lands where mill stones are obtained' and with a quarry nearby this is clearly the correct meaning. Ridley comes from *ridd leah* or 'the ridded woodland clearing'. Bearl is seen as Berehill in 1239, which is from Old English *bere hyll* or 'the hill where barley is grown'. Stalling is easily defined as 'the clearing with a fold for cattle'. Styford is recorded as Styfford in 1210 and as Stiford in 1262, derived from Old English *stig ford* meaning 'the path to the ford'.

Bywell Castle overlooks the river. Begun in 1430 it was never completed. All that remains today is a curtain wall and a still impressive gatehouse. In September 1878 the name of Bywell Castle was associated with the loss of seven hundred lives. An accident on the River Thames saw a collier named after this castle running into the *SS Princess Alice*, resulting in her sinking in minutes with the resulting fatalities.

C

Caistron

Found as Cars around 1160 and as Kerstirn in 1202, this is a combination of Middle English *kers* and Old Scandinavian *thyrnir* telling of 'the thornbush by the fen'.

Callaly

Here is a name listed as Calualea in 1161, this features Old English *calf leah* and describes 'the woodland clearing where calves graze'.

One local name worthy of note is Dancing Hall, not that any tango or military two-step was ever seen here. The place got its name for being where convicts arrested in the local area were held awaiting transportation to the magistrates at Powburn. Owing to the severity of sentencing during this period, the majority were facing the hangman's noose – hanging was also referred to as 'dancing on the hangman's rope'.

Callerton, Black

Listed as Calverdona in 1212, this comes from Old English *calf dun* and tells of 'the hill where calves graze'. The addition is self-explanatory, coming from *blaec* and distinguishing this from the following entry.

Callerton, High

A second 'hill where calves graze' from Old English *calf dun*, the self-explanatory addition distinguishes it from the previous name and is listed as Calverdon in 1242.

Cambo

This name comes from Old English *camb hoh*, listed as Camho in 1230, and describes 'the hill spur with a ridge'.

Possibly the most famous name in Cambo's history is Capability Brown. This famous landscape gardener of the eighteenth century attended the village school.

Cambois

Here is a Celtic place name meaning 'crooked' from *camm* and a reference to the shape of the bay here.

That coastal location is also seen in the name of the Buccaneer. Today this term is used as a synonym for 'pirate' and yet the two were markedly different. A pirate operated mainly in the Indian Ocean, a buccaneer was more organised, raided coastal towns rather than other ships, and operated in the Caribbean. Indeed the very word originates from the Arawak word *buccan*, a wooden frame used for smoking meat, most often manatee which inhabit the shallows of the Caribbean. This was then used for the French *boucan*, the frame used to smoke meat from feral cattle and pigs, and thereafter for those who hunted this game on the island of Hispaniola (modern Haiti and Dominican Republic) the *boucanier*.

Carham

Listed as Carrum in the eleventh century, this is Old English *carr* in a plural form telling of the '(place by) the rocks'.

Howburn is from *hoh burn* 'the stream in a valley'. Moneylaws describes the local area as having 'many hills'. Pressen is found as Presfen in 1176, this describes 'the fenland of the priests'.

Cartington

Records of this name include Cretenden in 1220. Here is 'the hill associated with a man called Certa, where Old English *ing dun* is preceded by a Saxon personal name.

Any image of Cragside House defines the name of a country house cut into the rocky hillside, although officially it takes the name of Cragend Hill which rises above it. Built in 1863, and subsequently extended, a hydraulic engine was installed five years later which took advantage of the water from the nearby lakes to power machines, particularly the laundry equipment. Two years later and the installation of a dynamo meant this qualified as the world's first hydroelectric power station, eventually bringing electric lights to this home.

Catcleugh

The earliest surviving record of this name is as Cattechlow in 1279. This is from Old English *catt cloh* and speaking of 'the ravine frequented by wild cats'.

Spithope, unchanged since the fourteenth century, describes 'the spit-shaped small valley'.

Catton

Listed as Catteden in 1229, this name comes from Old English *catt denu* and describes 'the valley frequented by wild cats'.

Leas Hall was constructed at the 'woodland clearings', the name coming from the plural of Old English *leah*.

Causey Park

Found as La Chauce in 1242, this is Middle English *cauce* which refers to 'the embankment, a raised way'.

Charlton (North & South)

A common place name, hence the expected additions, which is always Old English *ceorl tun* and 'the farmstead of the freemen or peasants'. Records of these names, which need no further explanation, appear as Charelton del North and Charelton del Suth in 1242.

Birchope would seem obvious as 'the valley where birch trees grow'. However the early forms seem to point to a Saxon personal name, giving something akin to 'the enclosed valley of a man called Beorhtwine'.

Chatton

Here is 'the farmstead of a man called Ceatta', where the Saxon personal name is followed by Old English *tun* and is recorded as Chetton in a document dating from 1178.

Allery Burn is a local stream from *alra burna* 'the stream of the alder trees'. The meaning of Bleaklaw depends entirely on whether the first element is Old English or Middle English, the former *blaec* with *hlaw* would give 'the dark hill or burial mound', while the latter *blake* with Old English *hlaw* would give 'the pale hill or burial mound'. The name of Brownridge explains itself, the colour refers to the soil.

Fowberry is from Old English *folan byrig*, easier to see in the thirteenth century records of Folebir and Follebiri, telling of 'the fortified place where foals are bred'. Hazelrigg comes from *haesel hrycg* or 'the ridge where hazel trees are found'. Hetton, found as the modern form as early as 1289, comes from *haeth tun* 'the farmstead on or by the heathland'. Lyham is found as Leyham in 1268, this represents *leah ham* or 'the homestead in the woodland clearing'. Weetwood is found as Wetewude in 1196, a name from Old English *waeta wudu* and easily seen as 'the wet wood'.

The Percy Arms recalls the family who were earls of Northumberland, undoubtedly the most powerful family in the north of England for most of the Middle Ages with the earldom first conferred upon them in 1766.

At nearby Ketley Crag is a rock overhang with excellent examples of prehistoric rock art in the form of cup and ring marks.

Cheswick

A name which can still be seen as coming from Old English *cese wic* and describing 'the farm where cheese is made'. This place is recorded as Chesewic in the early thirteenth century.

Chevington, West

Found as Chiuingtona in 1236, this is 'the farmstead associated with a man called Cifa', where the Saxon personal name is followed by Old English *ing tun*. The addition distinguishes it from a minor name.

Cheviot Hills

The earliest surviving record dates from 1181 as Chiuiet, later found as Chyvietismores in 1244 and as Chyviot in 1250. These forms show this cannot be of Saxon or Scandinavian origin, hence this name is certainly much earlier. There are five notable summits in the range, the Cheviot the highest at 2,674 feet, with Hedgehope Hill, Windy Gale, Cushat Law and Bloodybush Edge not far behind. With the exception of Windy Gayle, which straddles the border, all these are wholly within England and are part of the northern Cheviots, the southern range are lower.

Through these hills runs Broadstruthers Burn, a modern name thought to refer to 'the broad stong stream', while Harthope Burn tells of 'the stream in the valley frequented by harts or stags'. Harpath Sike is probably from *heriges paeth* 'the path or route of the army'.

Chillingham

Recorded as Cheulingeham in 1187, this name tells us it was 'the homestead of the family or followers of a man called Ceofel'. Here Old English *ing tun* follows a Saxon personal name.

Hepburn is found as Hybberndune in 1050, Hibburn in 1319, and Hilburn in 1346, this being from *hyll burna* 'the stream of the hill'.

Chillingham is famous for its castle and its herd of wild cattle. These have been penned in here since the Middle Ages. Under a hundred individuals mean the population has been inbreeding for many generations but continues to thrive. Mainly white, there is a little colouring around the feet, ears and noses, they stand out against the woodland pasture dominated by large oak trees. Whilst many believe this to be a glimpse of how livestock were kept in medieval England, these oaks were planted towards the end of the eighteenth century and the real ancient woodland is seen near the stream where coppiced alder trees are much older.

The claim this heard are directly descended from the primordial wild ox has little or no basis in known facts. However these wild animals exhibit behaviour of the ancestral wild cattle, although they are appreciably smaller than the extinct wild variety. Not in any way domesticated, the herd are much more aggressive than modern dairy and beef herds. On the 330 acres assigned to them the Chillingham herd show territorial behaviours, although two or three bulls tolerate one another in the smaller herds which overlap without any real conflict.

Chollerton

A name from Old Scandinavian *tun* and a Saxon personal name, this place is recorded as Cholucrton in the late twelfth century and was known as 'the farmstead of a man called Ceolferth'.

Broomhope comes from *brom hop* or 'the enclosed valley where broom grows'. An unusual name, Chipchase describes 'the hunting park associated with a man called Cippa'. From *cole denu* comes Cowden, 'the cool valley'. Felton Hill is from *feld tun* 'the farmstead by the open land', a name later also used for the hill.

Listed as Gunwarton in 1169, Gunnerton refers to 'the farmstead of a man called Gunnathor or of a woman called Gunnvthr'. Found as Rowchestre in 1348, Ruchester comes from Old English *ruh caester* and describes 'the Roman stronghold on the rough land'.

The church gate has a mounting block, steps made of stone to make it easier for parishioners to mount the horse or cart after leaving the church. Such were mainly for the use of the ladies, making mounting sidesaddle much easier. Mounting blocks were also positioned at the top of steep hills, allowing riders to remount after leading their horse up on foot.

Choppington

With an eleventh century listing of Cebbington, this is seen as coming from a Saxon personal name and Old English *ing tun*. Here the name originates from 'the farmstead associated with a man called Ceabba'.

As a pub name the Cherry Tree was first seen when it was taken from a nearby tree which acted as a marker. Soon afterwards it was used to represent a device found in the coat of arms of several families. The Half Moon Inn, which has given its name to Half Moon Street, is from an image found in numerous coats of arms. Although said to be 'half' it is actually a crescent moon. Either the 'horns' are pointing right or left or, in the case of the vast majority, straight up and an indication this family had ancestors who went on the Crusades to the Holy Land.

The sign hanging outside the Queens Head is sufficient to show this refers to the monarchy in general, rather than to one specific holder of the title. The Anglers Arms invite those who enjoy this sport to enjoy a drink and exchange stories of the one that got away. Similarly the Travellers Rest offers refreshment to passing trade which, as the age of the building shows, has done so for many years. The Shakespeare is a common pub name, indeed the Bard of Avon is second only to Horation Nelson in being the individual with the greatest number of premises named after him.

Colwell

Listed as Colewel in 1236, this is 'the cool spring or stream' and is derived from old English *col wella.*

Coquet, River

There is some doubt as to whether this name comes from the Old English or Old Scandinavian tongues, although in each case the meaning would be the same. It is not only the name of a river but also an island standing opposite the mouth of the island in the North Sea. However neither the water course nor the island have produced the name. This comes from the forest which grew in the valley through which the river flowed, thereafter transferred to the river and thence the island. Here Coquetdale is thought to describe the 'forest frequented by males of wild birds'.

The island is owned by the Duke of Northumberland and managed by the Royal Society for the Protection of Birds as an important site for seabird colonies. The most numerous are the 18,000 nesting pairs of puffins, however the most important would be the 92 breeding pairs of roseate tern, an endangered species which are more numerous here than anywhere else.

Corbridge

Records of this name begin around 1050 with Corebricg, the name describes 'the bridge near Corchester'. Here Old English *brycg* follows a familiar form of the Romano-British place name of Corstopitum nearby.

Aydon has been seen since the thirteenth century and probably represents *haeg denu* 'the valley where hay is obtained'. The suffix seen in Clarewood was not originally *wudu* but *worth* which, when prefixed by *cloefre*, describes 'the enclosure where clover grows'. Doepath Field began as *deop peth*, telling of the way through a hollow or valley'.

Meaning exactly what is says Gallow Hill is a fairly common name. From Old English *healh tun* comes Halton, 'the farmstead at a nook of land'. Howden describes 'the hollow in the valley'. Stagshaw is found as Stagschaue in 1296, this being 'the woodland enclosure where stags are seen'. Thornbrough, listed as Thorneburg in 1255 and as Thorneburi in 1262, this is 'the fortified place by the thorn bushes'.

While it would seen obvious the spring flower known as the bluebell is not the origin of the Blue Bell Inn, as there is just a single word for the flower, there are as many which give this as one word as there are two. However almost without exception the origin is not floral but religious, the bell a clear link to the church while blue has always been used to represent Christianity.

Pub names begin with the heraldic image seen outside the Black Bull Inn. Such a powerful image has been adopted by many families and thus difficult to know who. This is not the case with the Golden Lion, which is certainly representative of the Percys, dukes of Northumberland and the most powerful family associated with the county. A local family are remembered by the Errington Arms, an eighteenth century coaching house named after these former lords of the manor.

Established as a Roman fort in AD85, Corbridge archaeologists have uncovered many finds, the most notable being the Corbridge Lion and Corbridge Hoard. The sandstone carving of the lion was found at the beginning of the twentieth century. It will have been carved to adorn a tomb where, depicting the lion devouring its kill, it would have represented death's inevitable victory over life. The hoard comprised an iron-bound leather-covered wooden chest containing segments of armour, artillery bolts, a sword scabbard, carpentry tools, a wooden bucket, a tankard, a pulley-block and a lamp.

Cornhill-on-Tweed

Found as simply Cornehale in the twelfth century, this name comes from Old English *corn halh* and describes 'the nook of land frequented by cranes or herons'. The river name is a more recent addition meaning 'the strong one' and is discussed under its own entry.

Coupland

A name of Old Scandinavian origins in *kaupa land* and describing 'the purchased land'. This name is found in a document dated 1242 in exactly the form it appears today.

Cramlington

Found as Cramlingtuna in a document from the early twelfth century, here are four Old English elements *cran wella ing tun*, together they tell us this was 'the farmstead of the people resident by the stream of the cranes or herons'.

The Blagdon Arms is named for the Blagdon Estate at nearby Seaton Burn. The current mansion was built 1720-50, with later alterations. Constructed on the site of an earlier manor house, it was home to Matthew White, a local merchant and former Mayor of Newcastle.

The Plough has been popular since the sixteenth century, offering hospitality to those who worked the land which, at that time, was just about everyone. The Hind is probably a reminder of the hunt, in which case it could be said to be related to the Sportsman where any manner of sports could be played within or organised without.

The Bay Horse would be derived from a favourite animal, a bay does not refer to a specific brown colour but to any shade of brown horse with black points (ie mane and tail). A reminder of coal mining is found in the Brockwell Seam public house, while the Dudley takes a minor place name, one speaking of 'the *leah* or woodland clearing of a man called Dudda', and the Bridge stands near where the road crosses the local stream.

John the Clerk of Cramlington public house, Cramlington

In August 2012 the figure of a reclining woman was completed. Constructed from 1.5 million tonnes of earth it rises to 112 feet, is some 1,300 feet long, and covers 47 acres of public parkland. Perhaps this visitor attraction will suggest pub names of the future. While the official titles of Northumberlandia or the Lady of the North would be welcomed, it seems pub names would be more likely to use the colloquial name of Slag Alice.

Craster

A name speaking of 'the old fortification frequented by crows' from Old English *crawe ceaster* and appearing as Craucestre in 1242.

Locals enjoy a glass at the Jolly Fisherman. With the sea lapping the coast less than one hundred yards away it must always have had strong links to fishing. The additional 'Jolly' is commonly used to precede tradesman, indicating much merriment is to be had within.

The fishing theme is a reminder of Craster being known for its herring-curing. Craster kippers, also known as simply 'Crasters', is widely acclaimed as the finest kipper to be had anywhere in England and some claim it to be the birthplace of the kipper. It is one of the few foods to be named after its place of origin.

Cresswell

A name found in several counties in England, always from Old English *caerse wella* and describing 'the stream where watercress grows'. This name is recorded as Kereswell in 1234.

Snab Point is the best-known location here, somewhere far removed from where watercress grows. This sheltered bay is favoured by anglers in winter months for sizable cod move close to shore. Erosion of the low cliffs here has revealed fossilised trees and a small seam of coal. Whilst mining is not an option, coal can be gathered from the beach. This is sea coal, washed up when tides and winds are favourable and have left the beach a black colour.

Crookham

A name coming from Old Scandinavian *krokr* in a plural form, this is recorded as Crucum in 1244 and means the '(place at) the river bends'.

D

Dalton

Two places of this name in Northumberland, this example is near Hexham and is recorded as Dalton as early as 1256. This comes from Old English *dael tun* and describes 'the farmstead in the valley'.

Dalton

Another 'farmstead in the valley' from Old English *dael tun*, this example is found near Stamfordham. Again the earliest example, this time documented in 1201, is exactly the same as the modern form.

Dawdon Hall tells of its position, *dael dun* meaning 'the valley by the hill'.

Denwick

Records of this name include as Denewyc in 1242, this is from Old English *denu wic* and speaks of 'the specialised farm in the valley'. Almost all specialities would refer to dairy farming.

Minor names include Dunsheugh, the Saxon name followed by *hoh* referring to 'the spur of land of a man called Dunn'.

Detchant

Found as Dichende in 1166, the twelfth century record still says this is 'the end of the ditch or dyke' and is derived from Old English *dic ende*.

Dilston

The record of Deuelestune in 1172 shows this to be a combination of a Celtic river name and Old English *tun* referring to 'the farmstead by the dark stream'.

Diptonmill

Listed as Depeden in 1269, this features three Old English elements where *deop denu mill* tells us it was 'the deep valley with a mill'.

Ditchburn

From Old English *dic burna* and telling of 'the ditch stream', this name is recorded as Dicheburn in 1236.

Doddington

This is a name found as Dodinton in 1207, the name features a Saxon personal name and Old English *ing tun*. Here is 'the farmstead associated with a man called Dudda or Dodda'.

Ewart is a hamlet with a name which describes its location from *ea worth* or 'the enclosure by the rivers', the place is well isolated by the confluence of the rivers Glen and Till. Nesbit is from Middle English *nese bit* speaking of 'the piece of land shaped like a nose' - the feature is no longer visible so we must assume this was a human nose.

Downham

Found as Dunum in 1186, this is the plural of Old English *dun* and thus the '(place of) the hills'.

Duddo

Recorded as Dudehou in the early thirteenth century, here the Saxon personal name and Old English *hoh* combine to tell of 'the spur of land of a man called Dudda'.

Felkington is a local name found as Felkindon in 1237, this being 'the *tun* or farmstead of a the family or followers of Feoleca or Filica'.

A stone circle stands on top of a ridge north of Duddo. Known as Duddo Five Stones it was known as Duddo Four Stones until 1903 when the fifth large stone, which had lain flat for as long as records had been kept, was replaced.

Dunstan

The modern name is the perfect representation of Old English *dun stan* or 'the stone on a hill'. Indeed the earliest surviving record from 1242 also records the name as Dunstan.

Locally we find Mosscroft, not what it seems as early records of Musecroft in 1269 and Muscroft in 1323 show. Here the name comes from *mus croft* 'the small holding infested by mice', which of course begs the question as to why the inhabitants did not leave. Such derogatory names were not coined by the inhabitants, indeed few places would have been named by those who lived there, to them this was simply 'home' or 'our place'. Most place names are descriptions and probably began as simple instructions on where they could be found.

The Cottage Inn is a pub name which will have taken the name of the premises before it was ever considered as an inn.

E

Earle

From Old English *geard hyll* comes a name listed as Yherdhill in 1242. Here the name tells us this place was 'the hill with an enclosure'.

Earsdon

First seen in a document dated 1233 as Erdesdon, this is undoubtedly another combination of a Saxon personal name and Old English *dun* giving 'place at the hill of a man called Eanraed or Eoraed'.

Locally we find Brierdene, a name meaning 'the valley overgrown with briars'. Newsham is first seen in the twelfth century as Neusum, here two Old English words *niwe* and *hus* in a plural form *husum*, speak of 'the new houses'. Holywell is a common English place name, always referring to 'the holy spring'. Lysdon is recorded as Lidisdene in 1253 and as Lysden in 1533, this being 'the valley of a man called Lida of Hlyda'. Wheatridge is derived from Old English *hwit hrycg* and describes 'the white ridge of land'.

Easington

Found as Yessington in 1242, the suffix here certainly comes from Old English *tun*. However the first element is uncertain but generally thought to be an early name for the local stream.

Edderacres has been seen since the thirteenth century, telling of 'the *aecers* or lands of a man called Aethelred'. Flemingfield was named after the family of John le Fleming, who were certainly here by 1382. The first element of Haswell is clearly a personal name, probably a pet form and thus difficult to know if this is Hasso, Hesso, Hasse or Hesse with Old English *wielle* 'spring, stream'.

Hawthorn is from *haga thorn* referring to an abundance of 'hawthorn'. Monk Hesleden is found as Heseldene in the eleventh century, this seen as 'the valley where hazel grows belonging to the monastery at Durham'. Horden is probably *horh denu* 'the muddy valley' not *horh tun* 'the muddy farmstead'.

East Rainton

Listed as Reiningtone in the late twelfth century, here a Saxon personal name precedes Old English *ing tun* to speak of 'the farmstead associated with a man called Raegen or Regna'. The addition of East is to distinguish this from Yorkshire's Rainton and West Rainton in County Durham.

Edlingham

Listed as Eadwulfincham around the middle of the eleventh century, this comes from Old English *inga ham* with a Saxon personal name and tells us of 'the homestead of the family or followers of a man called Eadwulf'.

Bolton is a common name in the north of England, always from Old English for 'the farmstead with a (special) building'. Coe Burn is from the early regional term *cove* describing a stream which runs in a hollow, a cave, or similar recess. Learchild is a shortened form of 'the cold spring associated with a man called Leofhere.

Lemmington is from *hleomoc tun*, Old English for 'the farmstead where brook lime grows', more commonly known as speedwell a succulent herb with many uses. Roughley Wood takes its name from that part which contains no trees, for here is the *ruh leah* or 'rough woodland clearing'.

Eglingham

A name which is not only similar to the previous entry today but has a very similar meaning. Again listed in the middle of the eleventh century and with a Saxon personal name and Old English *inga ham*, this place appears as Ecgwulfincham, here is 'the homestead of the family or followers of a man called Ecgwulf'.

Harehope is seen as Harop in 1153, this is from Old English *har hop* 'the boundary in the valley'. Hedgeley is seen as Hiddesleie in 1150 and as Hiddesley in 1247, showing this is 'the *leah* or woodland clearing of a man called Hiddi'. Lilburn, found as Lilleburn in 1271, was then described as 'the stream of a man called Lilla'. Wooperton takes a Celtic term related to Welsh *gwy bre* and suffixes it with Old English *tun* to speak of 'the farmstead of or near the well-watered hill'.

Rimside Moor points to the topographical feature described as 'the edge of the hill'. From Old English *beo wic* comes the modern name of

Bewick, telling of 'the farm specialising in bee keeping', clearly honey production. Listed as Thitelittlenam, Tithlington and Titlinton in the latter half of the twelfth century, Titlington is from a Saxon personal name and Old English *inga tun* or 'the farmstead of the family or followers of a man called Titel or Tytel'.

Here the Tankerville Arms is a reminder of the earls of Tankerville. Members of the Grey family, the first creation saw the title bestowed upon John Grey, who fought at Agincourt and accompanied Henry V at the siege of Caen leading to him being granted the lordship (*seigneurie* in France) of Tilly in Normandy. After dealing with the Lollards in Powys he returned to France and granted the Comte de Tancarville. He was killed three years later at the Battle of Bauge near Anjou, one of many English nobility who fell owing to a lack of proper preparation and without the support of the archers which had made them such a powerful force.

Ellingham

Recorded as Ellingeham in 1130, this features a Saxon personal name and Old English *inga ham* and shows this to have begun life as 'the homestead of the family or followers of a man called Ella'.

Minor names begin with Doxford, a name with records surviving from the thirteenth century and telling of 'the ford of a man called Docc'. Hagg Wood has changed little from *hagg wudu*, an Old Scandinavian and Old English combination telling us it was 'the woodland copse which are regularly cut'. This is unusual, normally we would find the name reflecting the use of the material. Thus it is tempting to think this particular area was set aside to produce a product which required little volume but one with a regular and long-term demand - such as yew used to make bows.

From *hangra wielle hlaw* comes Hangwell Law, describing 'a spring emanating from an overhang'. Preston is a common name describing 'the farmstead of the priests', this example first seen in 1278 exactly as it appears today. Tinely, found as Tyndeley in 1278, seems to be from *tind hlaw*, quite literally 'the tine hill' and describing a region thought to resemble a harrow or a fork in shape.

The Pack Horse Inn takes us back to the days before the advent of the canal when the sole method of transporting any load load was on a pack horse. Be it corn, wool, even ore, it was slung across the backs of these sturdy animals, trains of which must have been an impressive and familiar sight across the land. Such treks would take in a number of inns, not only to take advantage of refreshment and rest for the horses and rider but to act as distribution and/or collection points.

Ellington

Despite the similarity with the previous trio of entries, this name may well have a different meaning. Listed as Elingtona in 1166, this could represent Old English *ael ing tun* and describe 'the farmstead associated with the catching of eels'. However it is also possible this represents a Saxon personal name and Old English *ing tun* giving 'the farmstead associated with a man called Ella or Eli'.

A local pub named the Plough Inn is a reminder of the days when virtually everyone scraped a living off the land. One of the most popular pub names since the sixteenth century, it offers hospitality to all with an instantly recognised image.

Elsdon

Found as Eledene in 1226, here is 'the valley of a man called Elli' where the Saxon personal name is joined by Old English *denu.*

Akenside is a minor name literally describing 'the oaken side', referring to the hillside where oak trees predominate. Carrick unites a Celtic word related to Welsh *caer* 'fort' and Old English *wic* seen as 'specialised farm'. Together these are understood as referring to 'the farm by or of the fort', there is no earthwork here today but may well have been in Saxon times. Chattlehope Burn is from a land feature, either 'the valley of a man called Cietel' or, from *cietel hop* 'the kettle-shaped valley'.

What began as 'the enclosed valley associated with a man called Cotten' is today marked on maps as Cottonshope. Davyshiel is first seen in the fourteenth century, all forms pointing to 'the shiel of a man called Davy (David)'. Earlside is an old place name, describing 'the seat of the earl'. Ellishaw comes from Old English *sceaga* and a personal name telling of 'the wood of a man called Illa'.

Garret Shiels is found as Gerardscheles in 1290, the name meaning 'the summer pastures of a man called Gerard'. Haining perfectly represents *haining,* very specifically 'the grass set aside for cattle', clearly excellent pasture which was protected by a hedge. Headshope features the suffix *hop*, telling us it was 'the enclosed valley of a man called Head'.

Heatherwick features the element *wic*, best defined as 'specialised farm' although almost always a dairy farm. However with the first element 'heather' this makes no sense and thus must be used as 'place among the heather'. With the Old English suffix *paeth*, Hudspeth speaks of 'the path of a man called Hod'. Monkridge holds no surprises in telling of 'the ridge of land of the monks'.

Ottercops is found as Altercopes in 1265, showing this originated in a Celtic term describing 'the summit of the battlements'. The corruption of the first element is down to the association with nearby Otterburn. From the dialect *hafre* comes Overacres, 'the land where oats are grown'. Raylees, first seen in the late fourteenth century, describes 'the clearing frequented by roe deer'.

Sting Head has the literal meaning of 'the clearing of the pole', which we understand as describing a place where pollarding encourages the growth of straight poles used for building fencing and walls, alternatively a place where such were stored and traded. Toft House is from Old Scandinavian *toft* already meaning 'the place of the house or building' before the addition of the comparatively recent 'House'. Yearhaugh comes from Old English *haugr yare* of 'the hill or mound by the fishery'.

On a hill overlooking Elsdon stands Winter's Gibbet, a replica of the original gibbet which stood here at Steng Cross. We need to journey back to 1791 when William Winter had returned to Britain after serving his sentence of transportation to Australia. Coming to Elsdon he learned of the kindness of an old woman named Margaret Crozier from two female faws (sellers of items in crockery and tin) named Jane and Eleanor Clark.

Together the three made their way to her secluded residence, bringing with them an ass to transport their ill-gotten gains. On the way they paused to rest at a sheep fold, it was here they were seen by a young shepherd boy - he was later to identify them and the butcher's knife with which he cut up the food they shared that day. Breaking into her home they murdered the poor woman and made off with the loot, a rather disappointing haul in their eyes. Tried and found guilty, William Winter was hanged at Westgate in Newcastle-upon-Tyne, his body then brought to Elsdon and hung in the gibbet which now bears his name.

Elwick

A twelfth century document gives this as Ellewich, showing this to feature a Saxon personal name with Old English *wic* or 'the dairy farm of a man called Ella'.

Old Scandinavian *brunnr toft* is seen in the name of the Bruntoft or 'the spring by the woodland clearing'.

Embleton

Recorded as Emlesdone in 1212, there are two possible meanings for this place name. Should this be a Saxon personal name and Old English *dun* this would give 'the hill of a man called Aemele'. However if this represents *emel dun* it should be understood as 'the hill infested by caterpillars'.

Broxfield represents *brocc* and *feld* or 'the open land frequented by badgers'. Craster is an abbreviated form of Old English *crawe ceaster*, speaking of 'the Roman stronghold frequented by crows'. Ewart's Hill describes 'the enclosure by the hill', although the modern name appears manorial. Falladon is from Old English *fealh dun* 'the ploughed land of or by the hill'. Heckley, found as Hecclive in 1250, is from *heah clif* 'the high cliff or bank'. Hefferlaw is a name describing 'the hill near the high ford', here 'high' is understood as meaning 'further upstream'.

Creighton Memorial Hall takes the name of Mandell Creighton. He served the community as vicar here 1875 to 1884, later becoming Bishop of London. Creighton would almost certainly have become Archbishop of Canterbury and leader of the Church of England had he not died at the age of 57. Stead Road is named after William Thomas Stead, an Embleton-born founder of investigative journalism and campaigner against child prostitution and child labour. His sphere of influence was enough to inspire George Bernard Shaw to write *Pygmalion*. He died in 1912, one of the many to lose their life on the maiden voyage of *RMS Titanic*. He is often cited as the most famous Englishman on board the huge vessel.

Eshott

Seen as Esseta in 1187, the meaning depends upon whether this represents Old English *aescet* or *aesc sceat*. The former would give the '(place at) the clump of ash trees', while the latter suggests 'the corner of land where ash trees grow'.

Etal

Listed as Ethale in 1232, this is derived from Old English *ete halh* and tells of 'the nook of land used for grazing'.

Perhaps that grazing land was given over to an animal which proved the inspiration for the naming of the local, the Black Bull. This is the only pub in the county boasting a thatched roof.

F

Falstone

Recorded as Faleston in 1255, as Faustan in 1371, Fauston in 1610, and Fawstons in 1663, this is a name meaning 'the fallow stone', a rock noticeably yellowish in colour.

The addition of 'moor' in Roughside Moor is quite recent, this began as Rughside in 1357 describing 'the edge of rough or unworkable land'. Emblehope has been seen since at least the fourteenth century, coming from Old English *emel hop* 'the valley of the caterpillar' it was probably used as a nickname to refer to its shape. Wellhaugh, a name describing 'the stream or string at the hill or mound', is recorded as Wellehawe in 1303.

Here we find Powtreuet, a name recorded as Poltrernerth in 1325. This name is clearly neither Old English nor Old Scandinavian. Hence unless this is an isolated example of a completely different tongue, this must come from the British or Celtic tongues. However as the only early example we have is comparatively late, we have no clue as to the origins.

The Pheasant Inn is probably chosen as it makes for an attractive image, although for the same reason it also made an excellent addition to a coat of arms.

Farne Islands

The earliest record of this place name dates from around AD700, however it is exactly the same as the modern form making it a candidate for one of the oldest unchanged place names in England. There was a Saxon word *fearn* meaning 'fern' which would fit the landscape, yet it seems likely this is of much earlier Celtic or even pre-Celtic origins.

Counting the number of islands depends upon the tide. At the highest tides there are just fifteen, while at their lowest ebb another five break the surface waters. There are two groups, the Inner Group and Outer Group. The former is comprised of Inner Farne, Knoxes Reef, East Wideopens and West Wideopens, all are united as a single land surface at very low tides. The Outer Group are formed by Staple Island, the Brownsman, North Warnes, South Warnes, Big Harcar, and the Longstone. Between the two is Staple Sound.

As with many isolated islands these became home to hermitages. Indeed it was as the home of Saint Aidan they are first recorded in 651. Two other monks later to be canonised, Saint Cuthbert and Saint Aethelwold, were resident here before the end of the seventh century. In 676 Saint Cuthbert introduced laws protecting the eider ducks and other seabirds nesting on his home. This legislation is thought to be the earliest protecting any bird species anywhere in the world. Doubtless these laws have helped to keep the Farne Islands an idyll for seabirds. Since records began in the eighteenth century, no less than 290 different bird species have been seen here.

While the last hermit, Thomas de Melsonby, died here in 1246, monastic cells were established soon after. These saw only two monks in permanent residence with a maximum of six at any one time travelling here from Durham. Three centuries later this cell was dissolved with King Henry VIII's Dissolution of the Monasteries in 1536 but remained a property of the Dean and Chapter of Durham Cathedral. Today they are held by the National Trust.

In 1838 the Longstone lighthouse beacon was unable to prevent the paddlesteamer *Forfarshire* from being wrecked on Harker Rock. During the night a few passengers managed to cling to the railings, including a Mrs Dawson still clutching the lifeless bodies of her two children. Next morning, still in appalling conditions, lighthouse-keeper William Darling and his 22-year-old daughter Grace rowed almost a mile in their 21 feet long coble to the wreck and rescued nine people from the wreck. More than forty perished, including the captain and his wife.

Felton

Listed as Feltona in 1167, this comes from Old English *feld tun* and tells of 'the farmstead in open land'.

The hamlet of Bockenfield has changed little since the first records from the thirteenth century, the name speaking of this being 'the open land near where the beech trees grow'. Glantlees features a first element which is unknown but seen as related to Teutonic *glint*, Old High German *glanz*, Old Swedish *glanta*, and Old Scandinavian *gletta* all of which mean 'bright, gleaming' or similar. Thus this would be 'the bright woodland clearing' when coupled with Old English *leah*. However we also know the hawk is known as *glente* in Old Danish and *glanta* in Old Swedish, so perhaps this is the meaning.

Helm stands on an obvious hill, one which appears to have been likened to 'the shape of a helmet' from Old English *helm*. Found as Overgares in 1255 and as Overisgar in 1250, Overgrass is from Old English *ufere gares* 'the upper triangular field'.

Two pubs here, the Northumberland Arms refers to the arms granted in 1951 to the county. The Stags Head has two equally plausible origins, either this is from an heraldic image or perhaps the huntsman brought back a trophy which was once displayed here.

Fenham

A name listed as Fennum around the end of the eleventh century, this is a plural form of Old English *fenn* and refers to the '(place of) the fenns'.

Fenrother

Here two Old English elements, *fin rother*, refer to 'the clearing by the mound or heap'. This name is found as Finrode in 1189.

Locally we find Gorfen Letch, from Old English *gorst fenn* or 'the fenland marked by furze' or, if the first element be a personal name, the 'fenland of a man called Gormund or Gornoth'. Heron's Close is named after a former owner, William Heron here before 1255 when the place name is first recorded. The surname comes from Old French *hairon* and means 'heron', probably originating as a nickname.

Fenton

The earliest record of this name is from 1242 where it appears exactly as it does today. Meaning 'the farmstead in a fen', this is from Old English *fenn tun*.

Fenwick

One of two places of this name in Northumberland, this example found near Kyloe. Here Old English *fenn wic* describes 'the dairy farm near the fen or marsh', the name recorded as Fenwic in 1208.

Fenwick

A second example of this place name, this found near Stamfordham and documented as Fenwic in 1242. Again this is from Old English *fenn wic* and describes 'the dairy farm by the fen or marsh'.

Flodden

Recorded as Floddoun in 1517, the name is derived from Old English *floh dun* and tells of 'the hill with stones'.

Flotterton

With a twelfth century listing of Flotweyton, this comes from Old English *flot weg tun* and describes 'the farmstead by the way that floods'.

The name of Plainfield appears to be a modern error for all early forms show this to be Fleinnfield. Thus there are two possible explanations, either 'the *feld* or open land of a man called Dleinn' or 'the arrow-shaped open land'.

Font, River

This small tributary of the Wansbeck joins the main river at Mitford. Recorded as Funt in 1251 and as Font for the first time in 1261, this comes from Old English *funta* or *font*, which is easily seen as 'fountain' and refers to the source of a lively spring.

Ford

No surprises for a name from Old English *ford* recorded as Forda in a document dated 1224 and meaning the '(place at) the ford'. Indeed, the only surprise is that it has never had a second element. The ford is a shallow crossing of the River Til, thought to have been used by monks and nuns travelling between the monasteries at Iona and Lindisfarne.

Crookham refers to 'the homestead by the crooked place'. Heatherslaw features the element *hlaw* here probably used to describe 'the hill of a man called Heathhere' rather than a burial mound. What began life as 'the farmstead of a man called Cynemaer' is now seen as Kimmerston.

Fourstones

From Old English *feower stan* and telling of 'the four stones', this name is recorded as Fourstayns in 1236. This refers to a tetralith here.

Framlington, Long

A name derived from Old English *ing tun* and a Saxon personal name. Here the name is found as Fremelintun in 1166, the addition is from much later and describes the shape of the village when viewed from the main road approaching the place, and tells of 'the farmstead associated with a man called Framela'.

Earl forms of Snook Bank include Schakelzerdesnoke in 1264, as Skalkelyerdesnoke in 1273, and as Snukbank in 1702. These earlier records seem to suggest the name means 'the snoke of a man called Skokull', not a Saxon name but one which is recorded among the Old Scandinavian speakers and used as a nickname from *skokull*, the poles of a cart between which the horse or oxen is tethered. Perhaps the Saxon personal name Shakel features, in which case this would describe 'the snook by the yard of a man called Shakel'. Yet if this is the early dialect word *shackle* this would speak of a place where cattle are tethered. However the present name only contains Middle English *snoke* or 'projecting piece of land'.

G

Gateshead

The very earliest known book to examine the origins of place names in England was also the first to examine the history of the nation and penned by the Venerable Bede in the eighth century. In his book Bede gives Gateshead as *ad caput capra*, which is literally 'at the she goat's head' in Latin. This appears to be an attempt to explain what, to him, was an unintelligible Celtic place name.

Perhaps Gateshead is the Gabrosenti, in which case it is easy to see how the British *gabro* was identified as being Latin *capro*. Thus with popular opinion swayed by Bede's work, the community became known as *gate heafod* 'the way at or to the headland'. This shows those who translated Bede's work had no notion of the developing

Once universally loathed, the Angel of the North has soon become a part of the scenery

English name. Pelaw is recorded as Pellowe in 1242, this representing a Saxon personal name and Old English *hlaw* telling of 'the mound or hill of a man called Peola'.

Local names include Bensham, where early forms are few but may point to a meaning of 'the homestead of a man called Beorn'. An intriguing definition is offered for Darncrook in 'the secret, hidden or remote crook of land'. Listed as Kibleswrthe in a document of 1185, the name of Kibblesworth features a Saxon personal name with the suffix *worth* and tells of 'the enclosure of a man called Cybbel'. Lamesley appears as Lamelay in 1297, from Old English *lamb leah* or 'the woodland clearing where lambs are reared'.

Gayton

Listed as Gaiton in 1162, this is derived from a Saxon personal name and Old English *tun* referring to 'the farmstead of a man called Gaga'.

Gilsland

A name meaning 'the cultivated land of a man called Gille or Gilli', where Old English *land* follows an Old Irish or Old Scandinavian personal name. This name appears as Gillesland in the twelfth century.

Locally the name of Wardrew is probably of Celtic origin. Records of the name begin as the modern form in 1479, however the name is certainly much older. The only way to discover the meaning is to compare this to similar names, the nearby Cumrew shares the suffix. If this is related to Welsh *rhiw* 'hill, ascent', then perhaps the first element is Old English *weard* 'lookout, protection'.

Glanton

A name from Old English *glente dun* which speaks of 'the hill frequented by birds of prey', although sometimes interpreted as 'the hill used as a look out place'. This name is documented as Glentendon in 1186.

Glen, River

Found in the eighth century writings of the Venerable Bede as Gleni, this is a Celtic river name related to Old Irish *glenn* or 'valley'.

Goswick

Listed as Gossewic in 1202, this is from Old English *gos wic* and refers to 'the specialised farm where geese are kept'. Clearly that speciality here are the geese.

Greenhaugh

Documented as Le Grenehalgh in 1326, this comes from Old English *grene halh* and describes 'the green nook of land'.

The Holly Bush Inn is most likely pointing to a nearby tree, used as a marker. An heraldic origin would be possible but only for leaves, branches or berries, the whole plant is unknown in heraldry.

Greenhead

From Old English *grene heafod* and listed as Le Greneheued in 1290, here is 'the green hill'.

Farglow is a local name which appears to be a Middle English mispronunciation, believing it to describe some picturesque view, of an earlier Celtic name.

The main road through Greenhead is Military Road, although designated the B6318. Its name, coupled with its long straight stretch, and proximity to Hadrian's Wall, have led to the belief this was a Roman road. However the road is recorded as being constructed by Hanoverian forces in 1746, designed to move troops quickly and efficiently in order to combat the threat of attack by the Jacobites from the north.

Greystead

Here the first element has nothing to do with colour, here the term means 'boundary' and is the origin for our modern phrase of 'grey area' when describing something uncertain or unclassifiable. With the suffix *stede* this name is a lasting reminder of 'the enclosed pasture at a boundary'.

Just to the southwest is Chirdon, where a Saxon personal name and Old English *denu* speak of 'the valley of a man called Ceorra'. Oxcleugh is recorded as Oxclow in 1279, this being 'the ravine by where oxen are reared. Snabdaugh is unchanged since the seventeenth century. From Old English *snab haugh* this is 'the enclosure by the small hill'.

Smales is found as Smale in 1279 and as Smalehopa in 1329, this might seem to be a personal name as it features a possessive 's' but there is no known personal name which fits the forms available. Thus the most likely origin is 'the small *hop* or valley', particularly as *hop* is already used to describe 'a small valley'.

Grindon

Although on the face of it different to the previous entry the meaning is exactly the same as 'the green hill', albeit the origin is a little different in Old English *grene dun*. The name appears as Grandon in a document from 1210.

Gunnerton

Found as Gunwarton in a document dated 1170, this seems to be from a Saxon personal name and Old English *tun* to tell of 'the farmstead of a woman called Gunnvor'.

Guyzance

Recorded as Gynis in 1242, this name is derived from the lords of the manor and remembers the 'family called Guines'.

The local name of Whirleyshaws is very difficult to define, despite the early records as Quirlecharr in 1350 and as Quarlecharr in 1356. One suggestion is *quarrel*, a dialect word modified from Scots *wharrel* 'a quarry', although the second element remains a mystery.

In the last months of World War II five 18-year-old soldiers died in the River Coquet in what became known as the Guyzance Tragedy. With the river in full flood their boat was swept over the weir and capsized while on a military exercise. Weighed down by their full combat gear all five drowned. Fifty years later, on 17th January, 1995, a service was held to mark the unveiling of a plaque to commemorate the five young soldiers.

H

Hadrian's Wall

Not only one of the most famous of our nation's heritage sites but one of the most famous Roman remains in the world. Hardly a schoolchild in the land does not know it is named after the Roman Emperor Hadrian who was head of the Empire when it was built inside four years, completed in AD126.

Correctly much of the wall was already there, the builders simply took advantage of the topography to emphasise the natural boundary between the two countries. While it could justifiably claim to be the most heavily defended border in the Roman Empire, although the extent of Roman influence elsewhere tended to stop at a natural boundary – usually water. The boundary was recorded by Rome as the

Hadrian's Wall

Aelian Wall, it is 73 miles long and is constructed from turf and stone at varying points along its route, depending upon materials available locally. With a width of ten to twenty feet and a height of twenty feet, together with the ditches, ramps and forts is still an impressive defensive feature. When manned and well-maintained it would have been an excellent deterrent.

Haggerston

Listed as Agardeston in 1196, this comes from an Old French surname and Old English *tun* and describes 'the farmstead of the family called Hogard'.

Hallington

Recorded as Halidene in 1247, here is 'the holy valley' from Old English *halg denu.*

Haltwhistle

Found as Hautwisel in 1240, this name comes from Old French *haut* and Old English *twisla* and describes 'the high confluence of two streams'.

Bellister features the suffix *ceaster*, the abbreviated pronunciation through Norman influence, and describes 'the Roman stronghold of a man called Bella'. Blenkinsopp has been recorded since the twelfth century, however the name meaning 'the *hope* or enclosed valley of a man called Blenkin' is much older. Caw Burn flows through the the place which gave it a name, 'the valley of a man called Cawa'. Another Saxon name is that of Coanwood or 'the wood of a man called Collan'.

Crook Burn almost explains itself as 'the twisting stream'. An Old Scandinavian place name, Farrow Shields is named for the '*sheles* or shed of a man called Freyvithr'. Featherstone, a name listed since the thirteenth century, is an unusual combination which has been argued as being 'rock with natural feather-like markings' or 'rocks which are as light as a feather', a definition should not be seen as literal but telling the rocks were not as heavy as they would appear.

Greenhead is self-explanatory and refers to the grassy high ground near Irthing. From Old English *healh strother* comes Haughstrother, a name meaning 'the marsh on or by the corner of land'. As noted

above: Haltwhistle proclaims itself the Centre of Britain
right: Twice Brewed, near Haltwhistle, Northumberland

elsewhere in these pages place names are coined by the neighbours, not the residents and Henshaw shows these were Christians for they describe this as 'the corner of land of the heathens'.

Melkridge comes from *meolc hrycg* 'the ridge of land producing good milk', a rather convoluted way of saying this place was renowned for its excellent pasture. Early records of Old Shield, such as Aldithescheles and Aldesheles, show the first element has been corrupted and mistaken for *ald* 'old' when the true origin is 'the summer pastures of a man called Ealdgys or Ealdwine'.

Plenmeller unites Old English *plaga* and the Celtic place name from *mel brez*, together these speak of 'the place of play at the bare or smooth-topped hill'. Redburn has two possible origins, either 'the red stream' from peat-stained waters, or Old English *hreod burna* 'the reedy stream'.

Sewing Shields is derived from 'the shielings or summer pastures associated with a man called Sigewine'. Bradley is a fairly common English place name, from *brad leah* it refers to 'the broad woodland clearing'. Found as Thurlewall in 1255 and as Thirlewalle in 1279, modern Thirlwall refers to a wall and the obvious one is that of Hadrian's Wall. The first element represents a dialect term *thirl* which means 'to pierce', meaning the name is understood as a place where people and goods could pass to and fro to the wall.

Thorngrafton features three Old English elements, *thorn graf tun* speaking of 'the farmstead of or by the thorny grove of trees'. Todhill is from Old English *tod hol* 'the place of fox holes'. Willimontswyke has been recorded since the thirteenth century, a name describing 'the specialised farmstead of a man called Willimot'. Wydon is found as Wyden in 1428, this being 'the wide valley'.

Most pub names referring to animal colouration invariably refer to an heraldic image. However in the case of the Spotted Cow Inn it seems more likely to be describing the local fauna, perhaps even one particular favourite animal. The Milecastle Inn takes the name of the Roman post of Hadrians Wall. The New Inn is one of the many pubs of this name which should be described as 'newer'.

Haltwhistle is one of two places claiming to be the geographical centre of Britain, the second is Dunsop Bridge in Lancashire 71 miles to the south. This surprising difference is down to the criteria used when taking the measurements. Drawing lines from the four cardinal points we do arrive at Haltwhistle, but only if the Orkneys are considered the northernmost point. This has not prevented the town having a Centre of Britain Hotel.

Harbottle

Listed as Hirbotle around the early thirteenth century, this is derived from Old English *hyra bothl* and is 'the building of the hireling'.

Hopperclose is found from the fourteenth century, the name describing 'the holding of a family called Hopper', the earliest known record of Hopper as a surname. Puncherton takes the name of the Norman lord of the manor, the Punchardon family hailing from Pontchardon in Normandy. The Drake Stone is an ancient landmark, a massive sandstone boulder once thought to be the source of supernatural powers. The reason for its name is unclear, it may be a landowner or former resident, although a corruption of Dragon Stone is also possible.

Harlow Hill

Old English *here hlaw* talks of 'the hill or mound associated with an army', most likely a Viking army as this is a Saxon name. The name is recorded in 1242 as Hirlawe.

Hartburn

The record of Herteburne in 1198 shows this to be from Old English *heorot burna* and describes 'the stream frequented by harts or stags'.

Angerton has few early forms, making it difficult to see if this is 'the *tun* or farmstead of a man called Asgeirr' or from *angar tun* 'the farmstead by the grasslands'. While *angar* is unknown in Old English, it is well-known in Old High German and would have been understood by the Saxons. Deanham is from *denu ham*, Old English for 'the valley with a homestead'.

Hartburn Sign

Seen since the late twelfth century, Hartington is from Old English *heorot weg tun* 'the farmstead near the path used by stags or harts'. Harwood House was constructed at 'the boundary wood'. Highlaws is from *heah hlaw* meaning 'the high hill'. Almost every record of Green Leighton shows the suffix here is not *tun* but *dun* thus this is 'the hill of a man called Lihtwine'.

Shaftoe is from Old English *sceaft hoh* or 'the shaft-shaped spur of land', the shaft in question being that of a spear. The earliest record of Fiselby dates from 1319 and is exactly as the modern form. This seems to be a German name with Old English *byge* meaning 'the place at the bend or curve of a man called Fisel or Fiso'. Wallington is derived from *wahl inga tun* or 'the farmstead of the family or followers of the foreigner'. Here *wahl* is Old English referring to the British or Celts who, at least in Saxon terms, were foreigners. Indeed this is the origin of Welsh and Wales.

Hartford, East

Found as Hertford in 1198, this is derived from Old English *heorot ford* and tells of 'the ford frequented by harts or stags'.

Hartley

Documented as Hertelawa in 1167, this shows an Old English origin of *heorot hlaw* and speaks of 'the mound or hill where harts or stags are seen'.

Hartleyburn

Listed as Hartelyburne in 1479, it is easy to see the three elements which comprise this Old English name. Here *heorot leah burna* speak of 'the stream by the woodland clearing where harts or stags are seen'.

Minor names include Glendue, a very accurate description of 'the black glen' for until comparatively recently this narrow feature was completely overgrown with trees making it well-nigh impassable.

Hauxley

A name either derived from Old English *hafoc hlaw* and 'the mound of the hawk' or the first element is a Saxon personal name and thus 'the mound of a man called Hafoc'. This place name is recorded as Hauekeslaw in 1204.

Hawkhill

Recorded as Hauechil in 1178, this is derived from *hafoc hyll* and 'the hill frequented by hawks'.

Haydon Bridge

Listed as Hayden in 1236, here is 'the valley where hay is made' from Old English *heg denu*.

Chesterwood unites two common Old English words where *caester wudu* describes 'the wood by the Roman stronghold'. Elrington has two possible origins, most likely 'the *tun* or farmstead of a man called Aelfhere' although 'the farmstead near the place where elder trees grow' cannot be ruled out.

Harsondale appears as Harestanesden in a document dated 1255, this being *hare stand dun* 'the boundary stone of the hill'. Records of Light Birks include Littelbirkes in 1296, Lythbirkes in 1328, and Lightbyrkes in 1368. Here the earliest record is thought to be erroneous, for 'the little birches' is much less likely than 'the light-coloured birches'.

From 'the woodland associated with a man called Lippa' comes the modern name of Lipwood. Morralee, recorded as Moriley in 1279, comes from *morigan leage* speaking of 'the swampy woodland clearing'. Ratton Row is first found in a document dated 1257. This is another example of a derogatory name given by neighbours, literally 'rat row' it would describe a row of houses thought to be fit only for vermin. Vauce is a Norman French name, from *vaux* this is a plural form meaning 'the valleys'.

Staward, listed as Staworthe in 1271 and Stannord in 1279, is from Old English *stan worth* 'the enclosure on stony ground'. Tedcastle is recorded since the fourteenth century, the name meaning 'the castle of a man called Tada'. Threepwood, recorded since the fourteenth century, is a name found in several counties of northern England and always referring to a region near a boundary as 'the wood of disputed ownership'. Whinnetley is a difficult name, the first element perhaps related to thicket as *whin* in the sense *whinet* and a small clump of whin or gorse' with the suffix is certainly *leah* or 'woodland clearing'.

A local stream name, one the subject of great speculation, appears in a document from the twelfth century as Manuggawburn and today is known as Manywaygoburn. The modern form is merely a slightly different pronunciation of the earlier example, itself already so corrupted as to leave the origins of the stream name a complete mystery save for the suffix *burna*. Sadly without further examples this intriguing name will remain undefined.

One local pub is the General Havelock. Major General Sir Henry Havelock is associated with the Indian Rebellion of 1857, when he led a British force recapturing Cawnpore from the rebels. Having enlisted as a second-lieutenant at the age of twenty when his legal education was interrupted when he and his father fell out, his career ended just a few days after his success of Cawnpore with his death at 62. Three times he was to lead his men to Lucknow to relieve the siege there and twice he turned back to save the already fatigued men under his command from the ravages of disease. Ironically his death was brought about by the dysentery he developed shortly after entering Cawnpore.

Healey

Found in 1268 as Heley, this comes from Old English *heah leah* and describes 'the high woodland clearing'.

Heaton, Castle

From Old English *heah tun*, this is 'the high farmstead' and is recorded as Heton in 1183. The addition is a reference to a manor house or hall which was here by the medieval period.

Hebron

Listed as Heburn in 1242, this is *heah byrgen* and Old English for 'the high burial place'.

Fenrother is a difficult name to define for early records, such as Finrode in 1189 and Finrothe in 1232, seem to point to different origins. All we can be sure of is the first element is not *fen* but likely represents the Scandinavian personal name Finn. The suffix is either Old English *ryther* 'the clearing', Middle English *rode* 'cleared from weeds' or perhaps *rid* 'cleared land'. Tritlington is from a Saxon personal name and Old English *inga tun*, recorded as Tirlington in 1210 this name speaks of 'the farmstead of the family or followers of a man called Tyrhtel'.

Heddon, Black

Appearing as Hedon in 1271, this comes from Old English *haeth dun* and refers to 'the hill where heather grows'. The addition probably refers to the dark shape of that hill.

Heddon on the Wall

As with the previous name 'the hill where heather grows' from Old English *haeth dun* and recorded as Hedun in 1175. The addition tells us it is to be found alongside Hadrian's Wall.

Here we also find East Heddon and West Heddon, two places with self-explanatory additions to an element with a common origin and one which is very different to that of the name of the parish but obviously influenced by same. Here we find early records of Hidewine in 1177, Hiddewin in 1187, Hydewyn in 1250, and Hydewin in 1298 all of which are quite different from its namesake. This name comes from *hid wen* perhaps referring to 'the hide of land by the winding stream'.

Close House was built at a place which already had a name, Old French *clos* meaning 'enclosure'. Eachwick comes from *ece wic* which, if given literally, would tell of 'the eternal specialised farm'. Most often the speciality of the *wic* would be dairy produce, however here the two in combination would probably tell of a place on a very long-term lease, possibly permanent.

A combination of *hoh tun* is found all over England as Houghton, 'the farmstead on a spur of land'. The stream known as Swarden Burn is named after the land through which it flows, described as *sware dun* 'the hill with heavy or sluggish soil'.

Hedley on the Hill

A name found as simply Hedley in 1242, this is from Old English *haeth leah* and tells of 'the woodland clearing where heather grows'. The addition here is self-explanatory.

The local is the Feathers Inn, a name which is derived from the plume of three ostrich feathers representing the Prince of Wales. However this is a place which prides itself as much on its food as its drink and to advertise such the management have cleverly utilised the sign. Keeping the general shape of the feathers, the three have been replaced by, reading from left to right, a fork, spoon and knife. Such ingenuity and originality is laudible in times when breweries seem to be opting for signs carrying no image but just the name.

Henshaw

Recorded as Hedeneshalch in the twelfth century, this is from Old English *halh* and a Scandinavian personal name and speaks of 'the nook of land of a man called Hetin or Hethinn'.

Hepburn

From Old English *heope denu* and 'the valley where rose hips grow' or *heopa denu* and 'the valley where brambles grow'. The earliest record is Hybberndune, an eleventh century record where the additional *dun* also refers to the overlooking 'hill'.

Hepple

Listed as Hephal in 1205 this, as with the previous name, has two possible Old English origins. If this is *heope halh* it is 'the nook of land where rose hips grow' or if *heope halh* 'the nook of land where brambles grow'.

Hepscott

A name from Old English *cot* and a Saxon personal name speaking of 'the cottages of a man called Hebbi'. The name appears as Hebscot in a document from 1242.

Heugh

Recorded as Hou in 1279, this represents Old English *hoh* and describes the '(place at) the spur of land'.

Hexham

A place name found as Hagustaldes in 685, this is Old English *hagustald ham* and 'the homestead of the warrior'.

The street names of Hexham seem to have changed more than most towns of its size. Perhaps this is due to the topography, where much of the original settlement sits atop the escarpment which naturally limits the expansion and means necessary improvements and alterations inevitably wipe away something of the earlier town. Examining both current and earlier names provides a glimpse into the later history of a town or village.

What is now Robbs Car Park was previously Alma Court which, along with Alma Place, also lost, were named from an inn which had taken the name of a battle fought in the Crimea. Here, on 20th September, 1854, the first engagement of the war saw a combined Anglo-French force of more than 60,000 defeat little more than half that number of the Russian army. The battle took the name of the River Alma, in turn named from the Tatar word for 'apple'. Eastgate, literally the 'eastern route into the place', was previously Bondgate from *bondigata* and telling us this road led past 'the land of the bondsmen' or peasant farmers.

What was Carlisle Road has become renamed in several places, in particular Shaftoe Leazes. This is found in several places in this part of the country and always describes a long strip as 'the shaft-shaped ridge of land'. Fore Street speaks of itself as 'the front entrance' and should be understood as the main entry point, there once being a Back Lane, too. Pudding Mews, previously Golden Lion Lane, was probably named for the sticky morass of which the road was made. While this is often seen as mud, it could easily have been sewage or, particularly considering the use of 'pudding', the intestines of slaughtered animals. A similar evolution is seen in what was Half Moon Yard, again named after a pub, becoming the self-explanatory Slaughter Houses.

St Hilda's Road certainly fits with the nearby names, where St George's Road, St Andrew's Road, St Wilfrid's Road and St Cuthbert's Road link to Priestlands Crescent, Priestlands Avenue, and Priestlands Road. However St Hilda's Road was previously known as Hilda Street which may have been named from a resident rather than the saint. Intake, earlier given as Intack, is an early field name pointing to 'the land recently enclosed for farming'. Merry Leazes was probably a deliberate change, for the modern form suggests something pleasant while the earlier Mirrey from *miry* tells us it was known for being muddy.

From the Saxon name Eardwulf comes the shortened form seen in Ardley as 'the woodland clearing of a man called Earda'. Aydon Shiels would have begun as 'the sheds or huts associated with a man called Ealdwine'. Telling us it was once known as 'the row (of dwellings) associated with a man called Bacga', the hamlet is now marked on the map as Bagraw. Coastley has nothing to do with the shoreline, this is 'the *leah* or woodland clearing of a man called Cocc'.

Dipton comes from *deop dene* 'the deep valley'. Dotland features an

Hexham Abbey

Old Scandinavian personal name in 'the cultivated land of a man called Dot'. Eshells features the suffix *sheles*, here either following a personal name in 'the shed of a man called Asketin' or, if the first element is Old Scandinavian *eski*, 'the shed by the ash trees'.

Hackford belongs to that group of places where, simply be defining the name, we see an image of the place as it appeared in Saxon times. Here the first element is *heck*, a regional variation on *haecc* or 'hatch gate'. This is understood to refer to a gate downstream from the ford, one which prevented animals from either straying or being carried downstream when the river was in spate.

Greenridge is easily seen as 'the grassy ridge of land'. Ham Burn is simply enough, from *ham burna* it describes 'the stream by a homestead'. Harwood Shiel speaks of 'the summer pasture by the boundary wood'. Hotley was Holtolaye in 1296, this referring to 'the *leah* or woodland clearing of a man called Holte'. Langhope is from Old English *lang hop* 'the long enclosed valley'.

Found as Lilleswrth in 1233 and as Lilleswude in 1663, this is probably 'the woodland of a man named Lilla', although that early record does seem to be describing 'the *worth* or enclosure of a man called Lilla'.

Hexham street signs

Hallstile Bank, Hexham

St Mary's Wynd, Hexham

A Hexham eatery inspired by the name of the coffee house in television's *Blackadder*

Recorded as Linelis in 1251, Linnolds is 'the farm of a man called Linnel', itself a pet form of Linbeald. Nubbock was once known as Yokesley, a name meaning 'the woodland clearing of someone called Yoke'. This is not a true personal name but comes from dialect *yoke*, used to describe a 'spouse or bedfellow' and genderless. First seen as Nobbock-scheles in 1479, Nubbock quickly lost its suffix leaving a name which has defied all attempts to define it.

Riddlehamhope, seen as Redeleme in 1214 and Ridlam in 1338, here is 'the homestead of the ridded or cleared valley'. Rowley is an Old English name from *ruh leah* meaning 'the rough woodland clearing'. The stream known as the Rowley Burn takes its name from the place. A place name of Steel comes from the word *steel*, a term once used by Scots dialect speakers to refer to 'a long line of rocks projecting into the sea'. Yarridge is recorded as Jernerig in 1232 and Yarwrigg in 1298, a name from Old English *gearwe hrycg* 'the ridge where yarrow grass grows'.

The Station Inn was named to advertise refreshment for travellers arriving on the (then) new mode of transport. Similarly the Globe conveys the message that this place will open its doors to all.

Doubtless salmon would have been served at the Salmon Inn, which means this is an advertisement for what was on offer. Another advert is seen in the Tap and Spile, the two parts hammered into a wooden barrel to enable easy dispensing of the contents. The tap is clear, a very simple version of the modern household tap, the spile is less clear. Anyone who attempts to pour any liquid from a container will know if no allowance is made to allow air to enter and replace the volume of liquid drawn off causes a bubble to enter and the steady flow is interrupted, often with messy results. Hence the spile was hammered into the top of the barrel and removed to encourage a nice steady flow of the contents through the open tap, then replaced when the tap was closed.

Location was in mind when naming the Shiregate public house, the Dipton Mill Inn, and the Old Tannery. However the Heart of England pub is clearly many miles from the centre of the country, indeed the name has nothing to do with location but can be found at nearby Hexham Racecourse, where the most famous race is the Heart of All England Steeplechase and the Heart of All England Cup presented to the winner. Traditionally this recalls the time when James VI of Scotland rode south to become James I of England and thus unite the two kingdoms. It is said he reached Hexham and announced: "Verily this is the heart of all England".

Other names include the Hare and Hounds, a reminder of the once popular sport of hare coursing. The Grapes, an image associated with the production of wines. The Coach and Horses shows this was a staging post for the major means of land transport in times before the railways. The Rose and Crown is a patriotic name, rose for the nation and the crown its monarch. The Kings Head carries an image of Charles II, a reference to 1660 and the Restoration of the Monarchy. Common as the animal lends itself readily to anthropomorphisation, the Fox Inn is as popular as ever.

Holy Island

Listed as Halieland in 1195, here is Old English *halig eg land* and quite literally 'holy island'. It is, of course, also known as Lindisfarne, a name discussed under its own entry.

Holystone

Found as Halistan in 1242, this name really does describe 'the holy stone' and comes from Old English *halig stan*. The stone would have been where the New Testament was preached.

Kidland speaks of 'the land of a man called Cyda' and is recorded as Kideland in 1271. Linsheeles, found as Lynsheles in 1292, tells of 'the *linn* or pool by the shiels or huts of the summer pastures'. Barrow Law is a corruption of Middle English *brere* with Old English *hlaw*, thus referring to 'the hill where briars grow'. Fairhaugh has been recorded since at least the thirteenth century, this being 'the fair enclosure'. Wilkwood is undoubtedly a reference to 'the *wudu* or wood of a man called Willa'. Yardhope is still recognisable as 'the *hop* or valley with a yard or enclosure'.

Horncliffe

A name derived from Old English *horna clif*, this describes 'the slope of the horn shaped hill'. The name is found in 1210 as Hornecliff.

Horsley

Recorded as Horseley in 1242, this name comes from Old English *hors leah* or 'the woodland clearing where horses are kept'.

Horton

The name of Horton tells of 'the farmstead associated with harts or stags' from *heorot tun.*

The hamlet of Cowpen refers neither to cattle nor to the folds in which they may be penned. This represents the plural of Old Swedish *kupa* literally 'cup' or 'bowl' but also used to describe a cup-like depression, perhaps man-made saltpans. Bebside is recorded as Bibeshet in 1203, as Befeset in 1292, and as Behisset in 1296, this is 'the *sceat* or corner of land of a man called Bibba or Bebba'.

It seems Stickley would feature a personal name and Old English *leah*, yet there is no record of a Saxon name Sticca, nor any similar Scandinavian alternative. However in Iceland we do find a record of a name Stykkis and while Icelandic references are rare they are not unheard of. The Scandinavians who travelled to our shores were simply a different generation to those who first populated Iceland, thus it would be reasonable to expect to find the personal name here too.

Howick

From the end of the eleventh century comes the record of Hewic, with a second listing in 1230 as Hawic. This is Old English *heah wic* and 'the high specialised or dairy farm', a name most likely referring to importance and not elevation.

The two landmarks of the village are Howick Hall and Howick House. Whilst it is possible to confuse the names, the two cannot be more different. The former is a Georgian mansion house and the seat of prime minister Charles Grey, 2nd Earl Grey. It was here a Chinese mandarin produced a special blend of tea to suit the local water supply and today the blend is known as Earl Grey. Howick House is a Mesolithic site, discovered at the cliff face and found to be a substantial building for the period with post holes, charcoal stains, and remnants of fires. Within these hearths were found burnt fragments of nutshells and bone. When these were radiocarbon dated they showed these to date from around 9,600 years ago and, for a time at least, gave Howick House the status of Britain's first house. Some of the fire remains show roasting nuts, and in large quantities, was all they were lit for. It has been speculated Howick House, here for around one hundred years, was where hunter-gatherers brought great quantities of nuts to be roasted in order to preserve them. If this is so it seems likely Holwick House may have been occupied seasonally, although those opposing this idea point to the size of the post holes showing this was a very substantial building for a temporary encampment.

Humbleton

Found in 1170 as Hameldun, this record is a perfect representation of the Old English *hamel dun* which speaks of 'the crooked or scarred hill'.

Humshaugh

A name derived from a Saxon personal name and Old English *halh* telling us this began life as 'the nook of land of a man called Hun'. The earliest known record of this place is from 1279, when it is recorded as Hounshale.

Records of the hamlet of Chesters shows it has only retained the suffix *caester* a Saxon term meaning 'Roman stronghold', the idea it is plural is a modern error. Early forms, such as Scytlescester in the early twelfth century, show this should be 'the Roman Stronghold associated with a man called Scytel'.

The Crown Inn is a clear reference to the monarchy, the message one of patriotism.

I

Ilderton
A name recorded as Ildretona around the beginning of the twelfth century, this name is derived from Old English *hyldre tun* and speaks of 'the farmstead where elder trees grow'.

Listed as Russeden in 1250, Russhden in 1346, Rusden in 1428, Rossedoun in 1580, Rosdon in 1663, and Rosden in 1712, which shows how Old English *risc dene* has evolved to the modern form but still meaning 'the valley growing with rushes'.

Ingoe
Found as Hinghou in a document from 1229, this features two Old English elements where *ing hoh* describes 'the peak with or by the spur of land'.

Ingram
This Old English place name tells of 'the homestead with grassland' and is derived from *anger ham*. The name appears as Angerham in 1242.

Hartside is seen as Hertesheved in 1255, this referring to 'the headland of the hart or stag'. Seen as Reveley in the thirteenth century, Reaveley speaks of 'the woodland clearing of the reeve'. A reeve was the officer who was charged with supervising the lands on behalf of the lord of the manor.

Irthing, River
A river name recorded as Irthin in 1169, Irthing again in 1169, Erthina around 1200, and Erthing in 1279.

K

Kielder

Found as Keilder in 1326, this was taken from a Celtic river name speaking of 'the rapid stream'.

Both the stream mentioned in the place name and the reservoir it feeds make the name of the Anglers Arms the ideal choice. The reservoir, along with the forest the best-known features here, has swallowed up much of the old railway line east to Falstone. Remnants include the Station Garage, alongside the old station building, and a castellated viaduct near Kielder Castle. Both the lake and forest of this name are man-made, the former the largest in the United Kingdom in terms of volume and the latter the largest forest in Europe.

Kibblesworth

A village seen as Kibleswrthe in 1185, where the first element is a Saxon personal name with a suffix of *worth*. Together these tell of 'the enclosure of a man called Cybbel'.

Kilham

Documented as Killum in 1177, this comes from Old English *cyln* in a plural form telling of the '(place of) the kilns'.

Killingworth

Seen as Killingwrth in 1242, here is 'the enclosure associated with a man called Cybbel'. The Saxon personal name followed by Old English *ing* and *worth*.

Kimmerston

Records of this place begin with Kynemerston in 1244. Here a Saxon personal name and Old English *tun* tell of 'the farmstead of a man called Cynemaer'.

Kirkharle

Listed as simply Herle in 1177, by 1242 this had becomes Kyrkeherl. This is a combination of Old English *leah* and a Scandinavian personal name and tells us this was 'the woodland clearing of a man called Herela'.

To the north is a reminder of the earlier form of Kirkharle, before the addition. Here stands Little Harle Tower, an eighteenth century mansion built around the fifteenth century pele tower. Hawick comes from *haga wic* understood as 'the hedged place where haws abound'.

Kirkheaton

Found as Heton in 1242, this shows the name was originally from Old English *heah tun* or 'the high farmstead'. Later saw the addition of Old Scandinavian *kirkja* or 'church' to distinguish from other Heatons.

Backstonerigg is a local name referring to 'the ridge where backstones are found', such flat stones were heated and used in baking.

Kirklevington

Listed as Levetona in the Domesday record of 1086, this describes 'the farmstead on the River Leven'. The river name, defined under its own name, suffixed by Old English *tun*. Later Old Scandinavian *kirkja* was added to show this settlement by a river was marked by its church.

Kirknewton

Recorded as Niwentona in the twelfth century, here Old Scandinavian *kirkja* combines with Old English *niwe tun* and tells us it was 'the new farmstead with a church'. Clearly the church was not mentioned in the earliest record but added later. Similarly the place would not have been particularly 'new' by the time it had become a part of the place name and thus should probably be viewed as 'newer'.

Locally we find Howtel, a name featuring the suffix *halh* or 'nook of land of a man called Holt'. Kilham is a pointer to what you should expect to find at a place name telling us it is 'at the kilns'. Seen as Langeton in 1255, Lanton is *lang tun* 'the long farmstead' and showing this was aligned along the main road approaching this settlement.

Paston is found as Pachestenam and Palestun in the twelfth century, these describing 'the farmstead of a man called Paelloc'.

Kirkwhelpington

The earliest record of this name dates from 1176 as Welpinton, when the origins were simply a Saxon personal name and Old English *ing tun* and telling of 'the farmstead associated with a man called Hwelp'. Later we find the addition of Old Scandinavian *kirkja* or 'church'.

Capheaton is recorded as Manga Heaton in 1274, not seen as Capptheaton until 1454. The basic name comes from Old English *heah tun* 'the high farmstead', which should be understood as referring to importance, not height above sea level. Stature is also seen in Magna, Latin for 'great', and later again where Latin *caput*, abbreviated to *cap*, describing 'the head or chief (settlement)'.

Catcherside, listed as Calcherside in the thirteenth century and Calchersyde in the fourteenth century, represents Middle English describing 'the cold cheer hill'.

Sweethope Loughs are two freshwater lakes below the windfarm to the west. Renowned for their large rainbow trout, the name tells of the 'attractive little valley'. The name of Crookdean describes 'the twisting valley'. Fawns is a name normally only seen north of the border, where it is used to describe 'moorland ground with white patches'. Ray is recorded as Raye in 1300, the name is thought to be derived from the dialect *wray* meaning 'landmark', although just what that represented is a mystery.

Kirkwhelpington is one of the few places granted a market charter but has never had a market. Henry III granted the lord of the manor, Gilbert de Umfraville, such in the thirteenth century. He even funded the building of a bridge over the River Wansbeck to improve the route into the town. Yet for some unknown reason no market was ever held and the privilege was revoked.

Knaresdale

In 1254 this name appears as Knaresdal, a name from Old English *cnearr* and Old Scandinavian *dalr* giving 'the valley of the rugged rock'. The first element is also seen in the minor name of Knar, itself transferred to the stream called Kanr.

Barns is a name first seen in 1325 where it appears as Le Bernes, the meaning exactly what it seems the '(place of) the barns'. Lambley describes 'the woodland clearing where lambs are reared'. Barhaugh is a lasting reminder of 'the hill where barley is grown', although it should be noted the word for barley was also used as a generic term for all grain crops. Snope was originally Snowhope 'the small valley where snow lies longer'. A more recent name is Town Green, where Middle English *grene* 'village green' follows Old English *tun* 'farmstead'. Found as Williameston in 1257, Williamston describes 'the farm of a man called William'.

Kyloe

The earliest record of this place name comes from 1195 as Culeia. This early form shows the origin to be Old English *cu leah* and 'the woodland clearing where cows are kept'. That the modern form is so different is purely through Scandinavian influence in both spelling and pronunciation.

Berrington is a hamlet recorded since the thirteenth century, this speaks of 'the farmstead of the family or followers of a man called Baere'. Fenwick is undoubtedly from *fen wic* 'the specialised or dairy farm by the fenland'.

L

Lamesley

One of the names which, for the author, is a delight to define for it gives a snapshot of Saxon life in England. Listyed as Lamelay in a document dated 1297, this is derived from Old English *lamb leah* and referring to 'the woodland clearing where lambs are pastured'.

Langley

A common English place name from *lang leah* and speaking of 'the long woodland clearing'. Few places were named by the residents, to them it was simply 'our place' or 'home'. Hence names were created by the neighbours and in this case the woodland clearing is described as 'long' which tells us it was named by a community further along the main road through the village. We know the general directions, if the settlement, as those living at ninety degrees around would see a different view and have seen it as *brad leah* 'the broad woodland clearing' which would have developed as Bradley.

Charlaw Moor is an Old English place name referring to 'the *hlaw* or hill of a man called Ceorra'.

Lanton

Recorded as Langeton in 1242, here the name comes from Old English *lang tun* and tells of 'the long farmstead'. Not that this village would have particularly 'long', simply would have appeared such being aligned along the main thoroughfare rather than spreading outward.

Learmough (East & West)

Two places with a common origin in Old English *laefer mutha*, the additions are self-explanatory. Recorded as Leuremue in 1177, this was 'the mouth of the stream where rushes grow'.

Lesbury

Found as Lechesbiri at the end of the twelfth century, this is derived from Old English *laece burh*. The literal translation would be 'the fortified place of the leech', but probably seen in a more favourable light if said to refer to the local physician who would have used leeches as part of many treatments.

The Coach Inn leaves us in no doubt this was one of the stops on the network of the coaching era.

Lever (River)

Here is a Celtic or British river name which is thought to describe the flow as 'smooth'.

Lilburn

Here a Saxon personal name and Old English *burna* combine to tell us this was 'the stream or a man called Lilla'. The name has changed little since it was recorded as Lilleburn in 1170.

Lindisfarne

Also called Holy Island, a name discussed under its own entry, the earliest record of this name appears around the end of the seventh century as Lindisfarnae. This is almost certainly Old English *farena eg* with a second place name and telling us this was 'the island of those from Lindsey'. This is an ancient Lincolnshire district, itself referring to 'the dry ground in the marsh of the Lindes tribe', who in turn take their name from their association with the city of Lincoln.

Fenham is probably from *fenn ham* or 'the homestead at the fenland'. Goswick is easily seen as Old English *gos wic*, giving 'the farm specialising in the rearing of geese'.

Linshiels

Found as Lynsheles in 1292, this is derived from Old English *lind scela* and refers to 'the shielings where lime trees grow'.

Littlehoughton

Recorded as Houcton Parva in 1242, the basic place name comes from Old English *hoh tun* and refers to 'the farmstead at the spur or ridge of land'. The addition of 'little', seen in the thirteenth century as Latin *parva*, distinguishes this from Longhhoughton.

Longbenton

Listed as Bentune in 1190, it is difficult to see if the Old English original name is from *beonet tun* 'the farmstead where bent grass grows' or *bean tun* 'the farmstead where beans are grown'. Later, in 1256, we find the name given as Magna Beneton, *magna* is Latin for 'greater'. The addition of 'long' is not seen before the seventeenth century, although it is quite likely it was in use soon after the thirteenth century record.

Longhirst

Seen in 1200 as Langherst and derived from Old English *lang hyrst*, this name tells us of 'the long wooded hill'.

Longhorsley

It is not until comparatively recently that this name was found with the additional 'long', referring to its alignment along the main thoroughfare. The addition distinguishes this from other places known as 'the woodland clearing where horses were kept' from Old English *hors leah.*

Garretlee is first seen as Gerardesley in 1296, this undoubtedly 'the woodland clearing of a man called Gerard'. Todburn is recorded as Totborne in 1434 and Todbourne in 1663. Either this is a personal name and 'the stream of a man called Tota' or is from *tod burna* 'the stream frequented by foxes'. Found as Welden in a document dated 1250, Weldon comes from *wielle denu* and describes 'the spring in the valley'.

As a pub name the Shoulder of Mutton Inn is quite common and often attributed to the coaching era, when this cut of meat was a popular repast for travellers. However the term is also seen as a minor place name, describing a short raised ridge of land resembling this cut. It seems both explanations have an equal chance of being correct.

Northumberland's role in the Roman Empire was evidenced by the discovery of a hoard of coins by metal detectorists. The coins, 61 sestercii and 9 dupondii, dated from the reigns of the emperors Vespasian and Marcus Aurelius AD69 to 180, a period when the Antonine Wall, much further north than Hadrian's Wall, marked the northwestern limit of the great empire.

Longhoughton

The earliest surviving record of this name comes from 1242 as Houcton Magna, a name from Old English *hoh tun* and 'the farmstead at the spur or ridge or land'. Here the addition of 'long' distinguishes this from Littlehoughton, although the thirteenth century record shows this was earlier Latin *magna* or 'great'.

Longwitton

A name from Old English *wudu tun* or 'the farmstead by the wood', the name appears as Wttun in 1236. This early form was well before the addition of *lang* or 'long'.

Early forms of the local name of Whittonstone show this is a corruption, clearly influenced by the parish name, of *whet stones*, such invaluable for honing metal tools.

Lowick

Seen as Lowich in 1181, this name comes from Old English describing 'the dairy farm on the River Low', the suffix *wic* being a specialised farm most often focusing on dairy produce. The River Low takes its name from *luh* or 'pool'.

Barmoor has records dating back to the thirteenth century, its name describing 'the moor of a man called Beaghere'. Gatherick is seen as Gateriswyk in 1281, which comes from *gaiter wic* and probably speaks of 'the specialised place of or marked by the wild dogwood tree'. Most often the speciality of the *wic* is dairy produce, however here this may not be the case. Holburn is fairly easily seen as 'the stream in a hollow', the name found as Hoburn in 1250.

Lucker

Found as Lucre in a document from 1170, this combines Old English *luh* and Old Scandinavian *kjarr* to describe 'the church by a pool'.

Lynemouth

A name speaking of the '(place at) the mouth of the River Lyne' which is recorded as Lynmuth in 1342. Here Old English *mutha* follows a Celtic stream name meaning simply 'stream'.

M

Matfen

A name recorded as Matefen in 1159, this is from a Saxon personal name and Old English *fenn* and describes 'the fen of a man called Matta'.

Meldon

The earliest record of this name is exactly as it appears today. Derived from Old English *mael dun* this speaks of 'the hill with a crucifix'.

Rivergreen describes 'the *hoh* or hook of land of the reeve', this name has been recorded since at least the thirteenth century.

Melkridge

Listed as Melkrige in 1279, this is from *meoluc brycg*, Old English for the ridge of land where milk is produced'. This name enables us to deduce a little more than a simple meaning, for it tells us of the excellent pasture which must have existed for some time in order for the name to stick.

Mickley

Found as Michelleie at the end of the twelfth century, this comes from Old English *micel leah* and tells this began as 'the large woodland clearing'.

What seem to be from Old English *ecg wielle*, Edgewell House is probably 'the spring on the edge', ie the side of a hill. However there is also the chance that the first element is a personal name, this would describe 'the spring of a man called Edge'.

Milbourne

Recorded as Meleburna in 1158, here is 'the mill stream' from Old English *myln burna*.

Mindrum

The mid-eleventh century record of Minethrum points to a Celtic origin of *monith drum* to refer to 'the mountain ridge'.

Mitford

Seen as Midford in 1196, this is 'the ford where two streams meet' from Old English *mythe ford*.

The local name of Aldworth, from *ealda worth*, describes 'the old enclosure'. Edington has three elements, a personal name with *inga tun* and telling of 'the farmstead of the family or followers of a man called Ida'. Fencewood can still be seen as referring to 'an enclosed wood'. Molesdon tells us this began as 'the *dun* or hill of a man called Moll'.

Nunriding Hall was built at the *nune ryding*, the area 'ridded or cleared by the nuns' of Holystone. The land was a gift to the nuns from Roger Betram when it was known as Baldwineswood. Isehaugh is seen as Ineshaulgh in 1370 and as the modern form from 1456, this being 'the enclosure of a man called Ine or Ini'.

Morpeth

A name from Old English *morth paeth* and recorded as Morthpath at the end of the twelfth century. This is something of a macabre memory of Northumberland history which must have been a well-known event at some stage, this 'path where a murder took place' recalls an event long forgotten.

Certainly the event predates the arrival of the De Merlay family, Norman lords of this manor from 1080-1266. During this time Ranulf de Merlay and his wife Juliana, herself from a powerful family as the daughter of Gospatric II, Earl of Lothian, founded Newminster Abbey. Earlier the family initiated the building of a classic Norman motte and bailey castle in Morpeth. Their nearly two centuries in Morpeth is not forgotten, De Merley Road is still named such more than eight hundred years after they departed.

The next family to take the reins in Morpeth are also seen in the locality. Greystoke is seen along Kings Avenue, they came here in 1266 and remained until 1487. Kings Avenue itself began as a surname, local families have also been remembered by names such as Sandersons Arcade and Tommy's Field.

Catch Burn has been seen since the thirteenth century, a name describing itself as 'the stream of a man called Caecca'. Cottingwood has three elements, telling of 'the wood of the family or followers of a man called Cotta'. Gubeon is a surname, the first record from the twelfth century is of William de Gobyon and one Hugh Gubion was sheriff of the county in 1296.

Morpeth sign

Shadfen is derived from Old English *sceald fen* and tells of 'the shallow fenland'. Shilvington is recorded as Schullington in 1250, as Schillingfon in 1316, and as Shilvyntoune in 1323. Here a Saxon personal name with Old English *inga tun* speaks of 'the farmstead of the family or followers of a man called Scylf or Scylfa'.

Woodside, near Morpeth's sign

Tranwell is found as Trennewell in 1267, here Old Scandinavian *trani* precedes Old English *wella*, which would describe 'the spring or stream where cranes or herons are seen' but here the first element seems to be used as a nickname. Early records of Ulgham point to an origin of 'the *ham* or homestead of a man called Ulf or Wulf'.

In order to attract customers public houses require a sign at the side of the road. Standing near the Great North Road one of the largest and longest-surviving signs gave a name to the Oak Inn. A name such as the Crown and Thistle is derived from symbolism, the 'crown' clearly refers to the monarchy and the 'thistle' shows this is the throne of Scotland.

Waterloo is a common enough name, referring to the final victory over the French under Napoleon. Heraldry is the basis for the sign of the Masons Arms, granted their coat of arms in 1473 the stonemason's skills among the most respected and admired throughout history. Workers in wood are represented by the Joiners Arms. A swan was a common device in coats of arms but was never given a colour, as in the White Swan here, until the discovery of the black variety, a species native to Australia. The Sun Inn offers a warm welcome, a 'place in the sun' seen as the ideal location.

The Anglers Arms offers hospitality to fisherman, a pointer to its whereabouts. Location is also the reason for the name of the Bridgend Tavern in Bridge Street, while the Cornerstone is found at the corner of the road, and the Railway welcomed those alighting at the station. The Wansbeck is named after the river, the Chevington from the place discussed under its own entry.

Murton

Found as Morton in 1204, this comes from Old English *mor tun* and describes 'the farmstead in moorland'.

N

Nesbit

Listed as Nesebit in 1242, here is a name from Old English *neosu hyht* which describes 'the promontory of the river bend'.

Netherton

A common place name, most often a minor name, always from Old English *neotherra tun* and dscribes 'the lower farmstead'. Around the middle of the eleventh century this name appears as Nedertun.

Newbiggin by the Sea

Another common name, particularly in the north of England, which unites Old English *niwe* and Middle English *bigging* telling of 'the new building'. The addition is self-explanatory and quite recent compared to the earliest known record of Niwebiginga in 1187.

The recent addition of 'by the Sea' will have advertised its location to prospective Victorian holidaymakers. Those seeking the sea air would not have come here in any great numbers before the advent of the railway, similarly the Railway Inn would have welcomed the weary travellers. While the Queen Victoria clearly honours the last monarch

above: Newbiggin-by-the-Sea's attractive welcoming sign
right: St Bartholomew's church, right on the coastline at Newbiggin-by-the-Sea

of the House of Hanover, the Queens Head is less obvious. However outside the latter is an image which shows this to be the next individual to be queen, although Queen Alexandra did not reign but was the popular consort of Edward VII.

The New Dolphin Inn is probably an image taken from the coat of arms of either the Fishmongers Company, or the Company of Watermen and Lightermen. The addition of 'new' informs us this is a more recent example of this name. The same is seen, along with a coastal reference, in the New Ship Inn. The Cresswell Arms is named from the nearby township, a name discussed under its own entry.

Newbottle

Documented as Beubotl in 1196, the Old English *niwe bothl* literally describes 'the new building' although should be understood as 'newer'.

Newbrough

Recorded as Nieweburc in 1203, this is from Old English *niwe burh* and tells of 'the new(er) fortified place'.

At Carraw we find a Celtic or British name, however it is difficult to know which language is the most closely related. This is likely *cadro*, which could be akin to Welsh *cader* 'chair' (often used to describe hills and peaks) or Old Breton *cadr* 'beautiful', with a suffix from Gaelic *reamhar* meaning 'fat, thick'.

Hawden comes from *haga denu* or 'the valley where haws are found'. Settling Stones features a term, as with saddling stone, to refer to a whetstone and something which must have been obtained here. The soil at Stonecroft cannot have been easy to work as it speaks of 'the stony small holding'. Wharmley comes from Old English *cweorn leage* 'the clearing of the mill'.

Newcastle-on-Tyne

There can be few, if any, towns and cities in England with a name with an obvious meaning. Here the *niwe castel* is first seen in a document dated 1130, although the listing is given in the Latin version of *Novem Castellum*. Of course this definition does raise the questions as to what happened to the original castle and when the newer version was constructed. To answer the second point first, this was built in 1080 by Robert, Duke of Normandy. Known by the nickname 'Curthose', he was the eldest son of William the Conqueror. There was no earlier 'castle' in the modern sense, this was simply an earlier defensive feature in the same place.

Here we find Benwell, a name first recorded as Bynnewalle in the eleventh century. If this record is correct this speaks of the place 'between the wall', quite plausible as this was the site of the Roman town of Condercum. However this is an isolated form and, as later records suggest something rather different, this isolated record could be erroneous, unfortunately there is no way of telling with the information to hand. Later records suggest this is describing the place as 'between the rivers', again quite correctly telling this the place where the Tyne is joined by its tributary.

As with any city there is a wealth of more recent history waiting to be tapped. A history largely overlooked until quite recently when the information staring us in the face was recognised, that being in the form of street names. Some of these are very old and represent minor place names. More recent names point to former residents and their families or even their occupations during the Middle Ages.

However the most interesting street names, and those most often overlooked, are the street names given by entrepreneurs and benefactors who helped develop our cities to accommodate the workers looking for jobs following the Industrial Revolution. While these streets named after themselves and their relatives may be seen as an ego trip and blatant nepotism today, perhaps this can be forgiven considering their contribution to the community.

Hence the names are separated into these various groups and begin with those which already existed as minor place names. Sandhill speaks for itself as a sandy hill leading down to the river. Likewise the street named The Side points to it being on the side of a hill close to the castle. One only has to look at The Close to see why the locals thought it was so narrow as to be described as 'close'. Similarly the narrowness of the stream gave a name to the Skinner Burn, although the only sign today is likely the street named Skinnerburn Road. Another stream, channelled to act as a gutter, almost certainly gave the name to Fenkle Street.

Forth Street and Forth Banks are derived from *fryhthe*, Old English for 'woodland scrub'. The term *laes* meant 'pastureland' and is the origin for the names Leazes Terrace, Leazes Lane and Leazes Park

Road. Bigg Market is named after the strain of barley which has four grains per ear and which was presumably sold here, as is the case with Cloth Market and Groat Market, the latter named for being where any grains were sold. It comes as a surprise to find Milk Market also sold meat as well as milk.

Many old towns and cities built on a hill have flights of stairs enabling pedestrians to move between the houses without having to follow the line of the streets. We still find these steps and their names, many require little explanation such as Castle Steps and Long Stairs. Breakneck Stairs, too, is easy to see as being exceptionally steep. Others take existing place names such as Dogleap Stairs, speaking of its twisted shape and, as the saying goes, as 'straight as a dog's hind leg'. Tuthill Stairs is similarly named for being where a lookout point had existed for millennia.

Until 1773 a white cross marked the site of the medieval market, the only marker today is Whitecross Way. The valley of the Lort Burn gave its name to Dean Street, *dene* being the Old English for 'valley'.

Bath Lane was named for the public baths built here in 1781. Dispensary Lane remembers the third location of the Newcastle Dispensary, where medical

Bigg Market, Newcastle-upon-Tyne

Clayton Street West, Newcastle-upon-Tyne

Dispensary Lane, Newcastle-upon-Tyne

Duke of Northumberland public house

treatment was offered before the National Health Service brought free health care to all. College Street marks the location of Durham University College of Medicine, built here in 1899 but subsequently moved. Drury Lane was named after that in London, both versions known for their theatres. Lombard Street was similarly inspired, both at one time being centres of business, with Silver Street showing where the silver merchants conducted their business.

Of course there are names which seem self-explanatory but these often require a little investigation. For example while South Street is named because it runs south from Forth Street, for the observer at the southern end of South Street it actually runs north to Forth Street. New Bridge Street can be followed to the bridge which was new in 1812. High Bridge and Low Bridge mark the site of the Upper Dean Bridge, which crossed the Lort Burn, long filled in.

Westgate Street does head west but refers to itself as being the 'way to the west' from *west geat* and nothing to do with a city gate unlike Newgate Street, which indeed led to a gate in the city wall but one which had not been 'new' since at least 1400. Gallowgate shares the same suffix but with a much more ominous prefix pointing to 'the way to the gallows'. Sandgate describes itself as 'the sandy road'. While nothing was documented to confirm, it seems the naming of Cross Street was suggested as it 'crosses' between the two roads at either end.

Religion has influenced names too, not least because of the power and landholdings of the church. The most obvious show the street leads to a particular church, as in the case of St Nicholas Street and St Nicholas Church and St Andrew Street from the church dedicated to St Andrew. St Thomas's Chapel left its mark in the naming of St Thomas's Street. St Mary's Place is from the former leper hospital, an establishment founded by the church of St Mary Magdelene. Once the clergy would lead a procession around the outside of the cathedral. Their prayers would end at the aptly named Amen Corner.

Orchard Street occupies land which had formerly been an orchard belonging to the local Carmelite friary. Low Friar Street was once known as Shod Friar Chare, 'shod' because the friars wore shoes and not barefoot as usual. High Friars was once home to the Grey Friars. Nun Street marks the site of the Benedictine Nunnery of St Bartholomew. Pilgrim Street was used by those travelling to the shrine of Our Lady of Jesmond and is one of the oldest streets in the city. Worswick Street takes the name of the Reverend James Worswick, founder of the first Catholic Church in the city following the Reformation.

Royals and the nobility are always a favourite choice and aside from the predictable King Street and Queen Street we also find names from the House of Hanover, rulers of Great Britain between 1714 and 1901, were the inspiration for Hanover Street and Hanover Square. Percy Street and Northumberland Street are named after the Percy family, dukes of Northumberland from 1766 who had their family seat at Alnwick Castle. Another of the family, Algernon, the 4th Duke of Northumberland, entered the House of Lords in 1816 as Lord Prudhoe, hence the name of Prudhoe Chare. Brunswick Place was named shortly after the marriage of the Prince Regent in 1795, his bride being Princess Caroline of Brunswick. Grey Street is named after the 2nd Earl Grey, who became prime minister of Britain in 1830.

Blenheim Street took the name of the battle won by the Duke of Marlborough, as is the family home of Blenheim Palace in Oxfordshire. Another battle, and an altogether more famous one, is remembered by Waterloo Street. Nelson Street is easily seen as named after Lord Nelson, best known for his role at Trafalgar where locally-born Cuthbert Collingwood, his second-in-command, was commemorated by Collingwood Street.

Friars Street, Newcastle-upon-Tyne

In any language this is Stowell Street, Newcastle-upon-Tyne

Marshalls Court, Newcastle-upon-Tyne

Nun Street, Newcastle-upon-Tyne

Pink Lane, Newcastle-upon-Tyne

The Battle of Otterburn saw the son of the Earl of Northumberland defending the land against raiding Scots led by James, Earl of Douglas in August 1388. The defender, known as Harry Hotspur, the attacker and the poem recording the events, *The Battle of Chevy Chase*, are marked by three street names: Hotspur Way, Douglas Way, and Chevy Chase.

Yet the author's favourite street names are those named for those who have worked to influence the development of a town or city. Sir Thomas Clavering, who served as MP for County Durham from 1768, gave his name to Clavering Place. Neville Street remembers the family who were earls of Westmorland, although not named directly but taken from Neville Tower which was named by the family. Another tower gave a name to Pink Lane. This was named from one of the towers on the city's wall, although just how Pink Tower got its name remains a mystery.

Berwick Street recalls Thomas Berwick, a local man made good who learned his trade through an apprenticeship and has become one of the best renowned Clayton Street was named to mark the work of former town clerk John Clayton, who died in 1890. William Scott was an eminent local judge who was made Baron Stowell and later had Stowell Street named after him. His brother John became Lord Eldon, hence the name of Eldon Square. Grainger Street and Grainger Market are after Richard Grainger who, prior to his death in 1861, built and planned much of the layout of the modern city centre.

John Blackett served four terms as mayor on the city in the eighteenth century, more than enough reason to name Blackett Street. Many members of the Ridley family have served the city as politicians, hence the naming of Ridley Place. Ellison Place remembers the family who have served the city as mayor and as sheriff since the sixteenth century. John Dobson Street is named after the man whose early nineteenth century architectural style can still be seen around the city. Property owner Sir George Saville is remembered by Saville Row and Saville Place.

Hood Street recalls John Hood who was the last mayor of the city before restructuring saw the creation of a different form of local government after the Municipal Corporations Act in 1835. Carliol Square recalls the family who were at their most powerful in the thirteenth and fourteenth centuries, several of whom held the office of bailiff and mayor. Plummer Chare is after Robert Plummer, a bailiff of Newcastle in 1378. Fenwick's Entry was once home to Alderman

Cuthbert Fenwick. Cocks Chare was named after Alderman Cocks who once lived here. However Mosley Street is named after Alderman Mosley, not his title for 'Alderman' was his Christian name. Few merit a street named after them more than Mark Akenside. Born here in 1721, he was made lame as an infant but went on to become a successful medical man and a poet.

Newburn-on-Tyne shares many similarities with the name of the city. Here the name is found as Nyweburne in 1175, Neuburne in 1203, and as Pipe Nieweburn in 1204. The addition of the 'pipe' shows this to be not only what it says as 'the new stream' but that it was a man-made channel. Dissington is found as Digentun in 1160 and as Dichintuna in 1190, a name from a Saxon personal name and Old English *inga tun* and telling of 'the farmstead of the family or followers of a man called Dica'.

Byker is a Middle English name which uses the established Old Scandinavian *kjarr* to describe the 'place by the *kjarr* or marshland'. Cowgate should not be taken literally for there was no gate, this pointed to an area where cows were allowed to wander on common land'. What began as 'the *wic* or specialised place of a man called Aelsige'. Gosforth is found from the twelfth century and comes from *gosa ford* or 'the ford of the geese'.

Black Blakehope is a contradiction, for the first and second elements mean quite the opposite. The *hop* or 'small valley' is joined with Middle English *blake* 'of pale colour' and related to modern 'bleak' rather than 'black'. The difference is entirely down to what is being described, it is the soil in the valley which is 'pale' and the whole valley 'shaded or dark'.

Heaton is a common name in England, nearly all represent *heah tun* 'the high farm'. While some will refer to elevation above sea level, most will describe importance or standing. Jesmond is found as Gesemue in 1204 and Jesemuth in 1242, this tells of 'the mouth of the Ouseburn'. The stream in question was formerly called the Yese, which explains the modern version much better than the local legend of 'the Jesus mound'. The stream name of Ouse Burn describing the 'yews' growing along the banks. Pandon is thought to be from *pampe dun*, which would give 'the thick-set or fat hill'.

Long Benton and Little Benton share an origin, however the record of Bentun in 1190 is insufficient to discern between *bene tun* 'the farmstead where beans are grown' or *beonet tun* 'the farmstead where bent grass is found'. Records of Dewley begin in the fourteenth century, this referring to 'the pasture where the dew falls' and must have been particularly heavy for it to become a part of its name.

Elswick eatures the suffix *wic* which here is understood as referring to a dwelling'. Listed as Elstwyc in 1189, as Alsistwic in 1198, and Alsiswic in 1203, this place began life as 'the dwelling of a man called Aelfsige'. Fawdon is found as Faughdon in 1309 and as Faudon in 1346, which is from Old English *fagh dun* 'the variegated hill', a reference to the colour of the land rather than the vegetation.

Killingworth is recorded as Killingwrth in 1242, a Saxon personal name followed by *ing worth* telling of 'the enclosure associated with a man called Cylla'. Longbenton is recorded as Bentune in 1190 and as Magna Beneton in 1256, either this is *beonet tun* 'the farmstead where bent grass grows' or *bean tun* 'the farmstead where beans are grown'. While the twelfth century addition was *magna* or 'large', the comparatively modern addition of 'long' refers to its alignment along the main road, thus appearing to be quite long when approached from either direction.

Kenton has been recorded since the thirteenth century, the name describing 'the farmstead

Thornton Street, Newcastle-upon-Tyne

St Andrews Street, formerly Darn Crook, takes the name of the oldest church in Newcastle-upon-Tyne

The former C.W.S Printing Works in Rutherford Street, Newcastle-upon-Tyne

The Herber Tower on Newcastle-upon-Tyne's Town Wall

The old Stoll Picture Theatre, Newcastle-upon-Tyne

of the king' from Old English *cyning tun*. Killingworth has hardly changed since its earliest known record from the thirteenth century, this being 'the enclosure of the family or followers of a man called Cylla'. Throckley is first found as Trokelawa in a document dated 1177, the name probably referring to 'the hill where beans are obtained' from Old English *throc hlaw*. With the earliest known listing from 1267 as Walkyr, Walker describes 'the *kjarr* or marsh overgrown with brushwood near the wall'. That wall is, of course, Roman.

Examining the public house names in Newcastle begin with the Black Bull. As stated elsewhere the vast majority of coloured animal names will be heraldic in origin. However in the case of the colour black there are over fifty different items following the colour and while the heraldic origin must remain likely, there are far too many examples for this to apply to all of them. It may be more than coincidental that the Gin Act came into force in 1736, when tavern owners covered their sign with a black velvet cloth and later added the colour to their name. It is impossible to know whether any specific examples can be explained by the Gin Act.

The Black Garter is unique and fits neither of these explanations. It is a combination of two other pub names: the Black Swan, itself created before the bird was found as native to Australia and designed to describe a real rarity; and the Star and Garter, symbolising the highest order of knighthood, the Most Noble Order of the Garter. Initially the combined pub became the Market Tavern but later changed, as a result of public pressure, to the Black Garter.

One of the most influential families in the northeast gave their name to the Blackett Arms. The title created in 1673 for William Blackett, Member of Parliament for the city, was later held by his descendants who were successful in shipping, lead and coal mining merchants. The family seat was Matfen Hall, Matfen which is now a hotel and country club. The Brandling Arms recalls the family who made their name as merchants, land owners and in coal, a family whose influence in the region can be traced back to the late fifteenth century. Similarly the Collingwood Arms recalls the work of the Collingwood family.

These goods will have been transported by boat and later railway, however earlier such heavy loads will have travelled by pack horse. Once great trains of these animals would have been a common sight, either travelling across country or waiting at one of the many stopping places on the network. One of which will have been a pub known as the Pack Horse, just as the example at Burnopfield. Such places will have offered rest and refreshment to both the animals and the men who led them. Other premises offered their hospitality to those on the road, hence the name of the Travellers Rest.

People are remembered in pub names too. Few individuals have more pubs named after them than the former solider and statesman Arthur Wellesley. Although the pub takes the name of his title, the Duke of Wellington. One person the duke would have met a number of times in his lifetime gave her name to the Queen Victoria.

The Bacchus also remembers someone, he being Bacchus the Roman god of wine. The Beehive is also symbolic, this image found in many coats of arms but also used as a pub name for the same reason, the beehive represents a busy environment and a busy landlord is a successful one and thus attracts even more business. The Blackie Boy, the local name for the Black Boy having been officially adopted, has been here since the eighteenth century on the site of a former blacksmith's shop (hence the name).

The Bob Trollop recalls a man more often referred to as Robert Trollope, a seventeenth century architect who worked in

Northumberland and neighbouring Durham. He designed the city's Guildhall, and many of the houses around those two counties. An heraldic sign, the Half Moon has nothing to do with our closest celestial neighbour. Originally the image of the 'half' moon was depicted as a crescent lying on its back with the two 'horns' pointing up. This image has been seen for more than four millennia and used by so many cultures and families it is almost impossible to determine the origin for the pub name.

One of the best-known characters in the history of the region is Harry Percy (1364-1403), a man known for gallantry and his fiery spirit. The eldest son of the 1st Earl of Northumberland, his story lives on in three of the plays of William Shakespeare and through his nickname used for the pub name the Hotspur and also for the Percy Arms. His family have given a name to the Northumberland Arms. The Newcastle Arms comes from the dukes of Newcastle, not the city itself.

A modern name in a Jacobean building, the Offshore 44 simply refers to the nearby shoreline, with the addition of the street number. The Waterline has similar origins.

The old County Court Building, Newcastle-upon-Tyne

The Old George Inn is the oldest pub in the city and boasts of visits from King Charles I. However the name refers, not to a king but, to the patron saint of England. The Pig and Whistle has been discussed for many years and without ever coming to a conclusion. Suggestions this was the traditional name for the bar on merchant navy vessels may well be true but still does not point to an explanation. Perhaps it is simply a created name but so long ago nobody remembers where it began, unlike the Pitcher and Piano which not only sounds like a pub name but also exhibits the alliteration which is the ad-man's dream. The same is true of the Quilted Camel.

The Rose and Crown is a common name and one which continues to be patriotic, the rose representing the nation and the crown its monarchy. One royal is seen in the name of the Three Bulls' Heads, part of the coat of arms of Anne Boleyn. Turnpike is a lasting reminder of the toll once charged to pass this point. The product is advertised by the Vineyard. Its location on one of the most famous ancient routes in the country by the Fosse Hotel.

The Victory is a different way of commemorating Lord Nelson, here by naming the pub after the admiral's flagship. The Fellsider speaks of those living on or near the fells, the Bay Horse an image feature a brown horse (of varying shades) with black mane and tail. An invitation is offered by the Traveller's Rest.

Newham

A name from Old English *niwe ham* which describes 'the new homestead'. This name is found as Neuham in 1242.

Newlands

The earliest surviving record dates from 1343 as Neuland. This is from Old English *niwe hus* or 'the new houses'.

Mereburn comes from *maer burna* and tells us it was 'the boundary stream'.

Newstead

Here is an Old English place name from *niwe stede* and describes 'the new settlement'. The place is documented as Novus Locus in the thirteenth century and as Newstede in 1339.

Newton-by-theSea

One of the most common place names in England means an addition is to be expected. From Old English *niwe tun* this describes 'the new farmstead' but should be understood as 'newer', ie replacing an earlier *tun*. Correctly there are two distinct parts to the place, Low Newton-by-the-Sea and High Newton-by-the-Sea, although the latter is something of a misnomer as it is half a mile away from the shore.

Here the addition speaks for itself and leads to the obvious origin of the Ship Inn. Nearby is the Blue Bell Inn, this is two words showing it does not refer to the spring woodland flower but the church – the bell link is obvious while blue is linked to Christianity. Note although many pubs will be found as one word (Bluebell), this rarely signifies the plant but is certainly influenced by same.

Ninebanks

Seen in 1228 as Ninebenkes, this is derived from Old English *nigon benc* and describes 'the nine banks or hillocks'.

Norham

Listed as Northham in the middle of the eleventh century, this comes from Old English *north ham* 'the northern homestead' or *north hamm* 'the northern hemmed in land'.

It is difficult to know if the first element of Buckton is *bucca*, which would give 'the farmstead where he-goats are reared' or, should this be a personal name, 'the farmstead of a man called Bucca'. Horncliffe is more likely to be 'the horn-shaped cliff' than 'the cliff on a horn or extremity of land'. Found as Ruleya in 1228 and coming from Old English *ruh leah* this name describes 'the rough woodland clearing'.

Shoresworth appears as Scoreswarthe in 1125 and as Sorwurth in 1203, this being 'the *worth* or enclosure of a man called Scorra'. Twizel comes from Old English *twislan*, telling us it is 'the place at the fork between two streams'.

Northumberland

The county name is recorded as Norhumberland in 1130, the early form telling of 'the territory of the Northhymbre'. Here the tribal reference is to 'those living north of the Humber', for in Saxons times their kingdom of Northumbria was much larger and stretched a considerable way south.

Northumberland

O

Ogle

This name comes from a Saxon personal name and Old English *hyll* to tell of 'the hill of a man named Ocga'. The name is first found as Hoggel in 1170.

Ogle Castle is correctly a fortified manor house, although it has been given Grade I listing. Humphrey de Hoggell was granted these lands by William the Conqueror shortly after the Conquest. Permission to crenellate was given to Robert Ogle in 1341, successive generations of the family holding the manor until 1597 when it passed by marriage to the Cavendish family and later to the Hollis family.

Ord, East

A name seen in 1196 as Horde is from Old English *ord* and describes 'the projecting ridge'. The addition is distinguish this from a similar minor place name.

The Salmon Inn is a reminder of how important the Tweed was, and indeed still is, as a source of salmon for the table. The reduction in cod numbers has generated more interest in other food fish. Farming has reduced the price of salmon considerably. Together these two factors have increased the amount of fish we eat in modern Britain. However, in comparison with our ancestors this amounts to little more than a drop in the ocean. Salmon, until recently considered a delicacy and a treat rarely seen on the dinner table, was once so plentiful that the peasants were sick and tired of having to eat its flesh. Indeed laws were passed forbidding masters from feeding their servants salmon on more than three days in a single week.

Otterburn

A name which still clearly speaks of 'the stream frequented by otters', this comes from Old English *oter burna* and is listed as Oterburn in 1217.

Potts Dultries is a local name found as Dortrees in 1275. The Potts family are recorded here around this time, the name telling us they were known as a source of 'trees for making doors', quite literally 'door posts'. Girsonfield is recorded as Grenesonefeld in 1331 and as Girsonfielde in 1586, this name describing 'the field of a man called Greenson'. Troughend features the Old Scandinavian element *trow* and speaks of 'the region where the troll dwells'.

Otterburn was the site of a particularly violent battle between the English and the Scots. In 1388 the English suffered a major beating when over a thousand were killed, captured or slain as they fled the scene, while Sir Henry Percy was captured. There can be few battles in history which have spawned ballads on both sides. The English heard the story through the lyric of *Chevy Chase* while the Scots version was entitled *Otterburn*.

Ovingham

Listed as Ovingeham in 1238, there are two very similar meanings for this name. Either this is a Saxon personal name and Old English *inga ham* and 'the homestead of the family or followers of a man called Ofa' or if *ing ham* would speak of 'the homestead associated with a man called Ofa'.

Ovingham's church of St Mary the Virgin

Harlow Hill comes from *here hlaw* or 'the hill associated with an army', probably the site of an encampment rather than a battle. From *hors leah* Horsley is easily seen as 'the woodland clearing where horses are reared'. Mickley represents Old English *micel leah* or 'the large woodland clearing'. Welton is seen as Waltenden in 1203 describes 'the *denu* or valley of a man called Wealta'.

Rudchester is recorded as Rucestre in 1250, Rodecastre in 1251, Rucestre in 1255, Rouecestre in 1268, Roucestre in 1296. With so many different contemporary late thirteenth century forms it seems there was as much confusion as to the real origin then as there is today. If the general theme can be trusted then this is undoubtedly 'the *caestre* or Roman stronghold of the red one'.

Two pubs here, both showing Ovingham's proximity to the river. The Bridge End Inn emphasises its location while the White Swan points to a common but most majestic sight on the water.

Ovington

Found as Ofingadun in 699, this name comes from a Saxon personal name and Old English *inga dun* and tells of 'the hill of the family or followers of a man called Ofa'.

The hamlet of Eltringham was once a separate settlement and one speaking of 'the homestead of the family or followers of a man called Heltor'.

P

Pegswood
Listed as Peggiswrth in 1242, this comes from a Saxon personal name and Old English *worth* and describes 'the enclosure of a man called Pecg'.

Pelaw
Recorded as Pellowe in 1242, this comes from a Saxon personal name and Old English *hlaw* and describes 'the mound or hill of a man called Peola'.

Pennines
The northern end of the backbone of England reaches the county, a range of hills which is probably linked to the British or Celtic *penn* 'hill' but how seems to be forever a mystery. There is also the possibility this was brought here by the Romans, the Apennines similarly forming the spine of Italy.

Pigdon
A name recorded as Pikedenn in 1205, this is derived from a Saxon personal name and Old English *denu* and tells of 'the valley of a man called Pica'. There is a second definition offered of 'the valley by the pointed hills' from Old English *pica denu*, yet none of these hills have significantly sharp summits.

Plenmeller
A name derived from the Celtic *blain mel brez* and describing 'the top of the bare hill', this is seen as Playnmelor in a document of 1279.

Ponteland

Found as Punteland in 1203, this is derived from a Celtic river name and Old English *ea land* and refers to 'the cultivated land by the river called Pont'. The river name itself is understood to refer to its position in 'the valley'.

From Old English *bean hrycg*, the hamlet of Benridge describes 'the ridge of land where beans are grown'. Another crop gave a name to Berwick Hill, this being 'the farm which specialises in barley'. What was recorded as Brinchelawa in 1177 is today seen as Brenkley and refers to 'the *hlaw* or hill of a man called Brynca'. Coldcoats is a common enough name, always referring to 'the cold or exposed cottages'.

Darras is a manorial name, this place was held by the Darrayns family, who hailed from Airaines (hence d'Airaines) in the Somme region of France. Dinnington features a Saxon personal name and Old English *inga tun* telling of 'the farmstead of the family or followers of a man called Dunna'.

Higham Dykes speaks of 'the embankments near the high homestead'. Kirkley is Old English for 'the woodland clearing with a church', with the first element showing clear Scandinavian influences. Milbourne is not difficult to see as the '(place at) the mill stream'. Mason is seen as Merdessen in 1296 and as Merdesfen in 1336, speaking of 'the fenland of a man called Merdo'.

A small castle-like building, largely destroyed by a Scottish army on a foray culminating in the Battle of Otterburn, has been used to produce what is now the Blackbird Inn. Such a name will be heraldic with the reference to 'a black bird' such as a crow, rook or raven and not to the blackbird. One room in the modern pub, known as the Tunnel Room, is said to be where a bricked up entrance behind the present-day fireplace hides a tunnel linking it to St Mary's Church.

Port Gate

Recorded as Portyate in 1269, this was originally from Old English *port* here used to mean 'gate' and a reference to the gap in Hadrian's Wall allowing Watling Street to pass through. Later this was suffixed by *geat* which is easily seen as meaning the same thing and thus the name can be defined as 'gate gate'.

Prendwick

Documented as Prendewic in 1242, this features a Saxon personal name and Old English *wic* and tells us it was 'the specialised farm of a man called Prend'. As stated elsewhere almost always this was specialising in dairy produce.

Preston

A common place name, most often found with a second element for distinction. All examples come from Old English *preost tun* to describe 'the farmstead of the priests', with this place listed exactly as it is today in 1242.

Prestwick

Found as Prestwic in 1242, this comes from Old English *preost wic* and describes 'the specialised (dairy) farmstead of the priests'.

Prudhoe

Listed as Prudho in 1173, this name features a Saxon personal name and Old English *hoh* referring to 'the hill spur of a man called Pruda'.

Locals enjoy a drink at the Adam & Eve public house, the sign features the image of the first man and woman under a tree, an image which is clearly taken from the story in *Genesis*. The sign painter was probably unaware the name is heraldic, aptly (considering their association with apples) adopted by the arms of the Fruiterers' Company.

The West Wylam Inn takes the nearby place name discussed under its own entry. Clearly the Fox and Hounds is a reference the now banned blood sport of fox hunting, although whether

Prudhoe welcomes us

this was a meeting place for same or where the hounds were kennelled is unclear.

Prudhoe Castle

R

Redesdale

Recorded as the modern form as early as 1075, this speaks of 'the valley of the River Rede'. The river name is of Old English origin, *reade* speaking of 'the red one', the colouration caused by the local soils.

Chesterhope has changed little from the Old English *caester hop* 'the valley associated with a Roman stronghold'. Corsenside links Old English *saete* and a Saxon personal name, giving the '(place of) the dwellers associated with a man called Crossan'.

The River Rede, which gave its name to Redesdale

The hamlet of Farnycleughcan still be seen as 'the valley where ferns abound'. Golden Pot is the name of an area which included two stone blocks, markers known as the Outer Golden Pot and Middle Golden Pot. The name refers to the hollowed out top, presumably to allow some offering to be left. This was not a precious metal but 'the pot of a man called Golda and (later) his sons'.

Hernehouse is a good description of 'the house in the corner of land'. Karswelleas shows Old Scandinavian influence, even though the origin is Old English in 'the spring or stream where cress grows' and would normally have become Cresswell. Learm is from *laeg ham* 'the fallow or unploughed land by the homestead'.

Rennington

Recorded as Reieneingtun in the twelfth century, here the name is derived from a Saxon personal name and Old English *ing tun* and describes 'the farmstead associated with a man called Regna'.

The Horse Shoes Inn will probably have used old shoes as its original sign, a clear indication to the largely illiterate traveller he could find both rest and sustenance for both himself and his steed.

Riding Mill

Listed as Ryding in 1262, here the Old English *ryding* is the origjn, a word meaning 'a clearing'.

Rochester

Found as Rucestr in 1242, the name is derived from Old English *ruh ceaster* and 'the rough Roman stronghold'. The stronghold spoken of was named by the Saxons. When the Roman station was here, built to defend the important Roman artery running from York, through Corbridge to Melrose and now known as Dere Street, it was known by them as Bremenium - a name meaning 'the place of the roaring stream', a rather literal description of a young, fast-flowing water course. The fort is almost square, 445 feet east to west and 485 along a north-south axis, giving a five acre site. There is evidence this stone construction replaced the earlier turf rampart during the period when Quintus Lollius Urbicus was Governor of Britain. He is recorded as staying here *en route* to Scotland to oversee the building of the Antonine Wall in AD142. Today the stone remains show these were appreciably heftier than other Roman forts and were equipped with catapult-like machines known as onagri, a smaller and faster-loading version of the better known ballista.

The first element of Bell Shiel is the personal name Bell or Bella, the suffix is Middle English *shele* meaning 'hut, shed'. However this is not seen before the seventeenth century, prior to that this was 'the *hop* or enclosed valley of a man called Bell or Bella'.

Rock

A name featuring Middle English *rokke* and recorded as Rok in 1242, the name describes the '(place at) the rock, peak'.

Roddam

Listed as Rodun in 1201 and coming from Old English *rodum*, the name can be defined as the '(place at) the clearings'.

Roseden

Seen as Risseden in 1242, this is 'the valley where rushes grow' and is derived from old English *rysc denu.*

Ross

A name found as Rosse in the early thirteenth century, but the name is much older and is derived from Celtic *ros* meaning 'hill spur'.

Rothbury

Recorded as Routhebiria around the beginning of the twelfth century, there are two possible origins for this place name. Either this is from a Saxon personal name and Old English *burh* and 'the stronghold of a man called Hrotha' or, should the first element by old Scandinavain *rauthr*, this is 'the red stronghold' and a reference to the colour of the rock hereabouts.

Bickerton is Old Scandinavian *bekkr* and Old English *tun* 'the farmstead by a stream'. Brockley Hall is a name recorded as Brockleygehirst in 1309, when the name clearly describes 'the clearing of or by the wooded hill frequented by badgers', since when the element *hyrst* has been dropped. Fallowless Burn is a stream name taken from the neighbouring land described as 'the woodland clearing of varied colours'.

Flotterton is from Old English *floten weg tun* 'the flooded way to the farm', almost certainly describing a road which was liable to flooding by the nearby Coquet. Seen as Heselhyrst in 1268, Hesleyhurst tells of 'the wooded hill where hazel grows'. Moralhirst comes from Old English *myrige hylde hyrst* or 'the woodland of the pleasant slope'. Pauperhaugh is seen as Papwirthhalgh in 1120 and Papwurthhalgh in 1250, showing this was originally 'the corner of land by the specialised (probably dairy) farm of a man called Pappa'.

Tosson is a difficult name to define, probably best explained as *tos vin* or 'the field where flax is grown'. Trewhitt is seen as Tyrewyt in 1229, a difficult name which is probably a personal name influenced by *tirwhit* a dialect word for 'lapwing'. Wreighill, found as Werghill in 1292, here is 'the hill of the felons' from *wearg hyll*. Such probably refers to where gallows were erected.

The Turks Head public house gets its name from the image found on the coat of arms of many families, adopted to show how their ancestors were proud to have fought on the side of Christianity in the Crusades. The Queens Head features a coin with the image of a young Queen Victoria.

As with many villages in the border region Rothbury has had a turbulent and bloody history. It is remembered as a rest stop for the Earl of Derwentwater and his followers in 1715. Few made the return journey having suffered a heavy defeat at the Battle of Preston. Nearby Lordenshaw Hill has the largest number of rock carvings in the county. Around one hundred have been uncovered, the simplest are cup-marked boulders although there are as many complex carved panels.

Rothley
From Old English *roth leah* and 'the wood with clearings', this name is recorded as Rotheley in 1233.

Ryal
Found in a document from 1242 as Ryhill, this shows the origin to be Old English *ryge hyll* or 'the hill where rye is grown'.

Ryhope
A name recorded in 1050 as Reofhoppas, this name has two possible Old English meanings. Either this is *hreof hop* 'the rough valleys', or *refa hop* 'the valley of the reeve'.

Ryle (Great & Little)
Two places with a common origin in Old English *ryge hyll*, here is 'the hill where rye is grown'. The additions are later and self-explanatory.

Ryton

Listed as Ritona in 1183, this common name is derived from old English *ryge* tun 'the farmstead where rye is grown'. It is unusual to find such a common name without a second distinguishing element.

Barlow is a minor name recorded as Berleia in 1380, showing this was not 'the hill where barley grows' but 'the open land where barley is cultivated'. Chopwell is a hamlet which began life as 'the *wielle* or spring of a man called Ceappa'. From *crawe krok* meaning 'the corner of land frequented by crows' comes Crawcrook, although the first element could perhaps be used as a personal name. Milkwell Burn has nothing to do with dairy farming, this describes the colour of the water in 'the milky waters of the stream'.

S

Sandhoe

Recorded as Sandho in 1225, this is from Old English *sand hoh* and describes 'the sandy hill spur'.

Scrainwood

Listed as Scravenwod in 1242, this is an Old English place name from *scrawa wudu* and telling of 'the woodland frequented by shrewmice or villains'.

Scremerston

Found in a document from 1196 as Schermereton, this features a Middle English surname and Old English *tun* and tells of 'the farmstead of the Skermer family'.

Local geology allowed many coal seams to develop, these two-feet thick have been dated and proven to be among the earliest ever found in Britain. This fuel source, together with the clay soil, allowed the manufacture of bricks and tiles alongside the lime kilns.

Seahouses

No doubting the origin of this place name. Doubtless when this was coined there were only a handful of buildings on the coast.

The coastal reference of the place name is mirrored in its public houses. In the case of the Schooner Inn the name refers to both the sea-going vessel and also refers to a glass in which sherry is traditionally served. The Olde Ship was not 'olde' until relicensed in the twentieth century, the original nineteenth century licence simply called it the Ship. Both vessels will find safe haven in the place which gave a name to the Harbour Inn.

Seaton Delaval

A name which means 'the farmstead by the sea' from Old English *sae tun*. The name is first found in 1270 as Seton de la Val, a reminder the de la Val family were here in the thirteenth century, their family name derived from Le Val in Normandy. The family still have connections to the area, their descendants still owning land here. Indeed the current Lord Hastings, 23rd of this line, was named Delaval.

A folksong entitled the *Blackleg Miner* tells a story of Seaton Delaval:

Oh, Delaval is a terrible place
They rub wet clay in the blackleg's face
And around the heaps they run a foot race
To catch the blackleg miner

So dinna gan near the Seghill mine
Across the way they stretch a line
To catch the throat and break the spine
Of the dirty blackleg miner!

Seghill

Records of this name begin with Syghal in 1198, this is 'the nook of land by the stream called Sige'. Here the origins are Old English, the river name meaning 'slow moving' with *halh.*

As a pub name the Plough offered a welcome to those who worked the land. A popular name since at least the sixteenth century when it was an invitation to those who worked the land.

Sharperton

This comes from Old English *scearp beorg tun* and refers to 'the farmstead by the steep hill'. The earliest known record is as Scharberton in 1242.

Shilbottle

Seen as Schipplingbothill in 1242, this is 'the building of those of Shipley' with Old English *inga botl* and the following place name.

Guyzance, recorded as Gsynes in 1240, came to Shilbottle as a family name. We are already aware of a number of surnames – Gisnes, Gysnes, Gynes, etc., – which are all traceable back to the town of Guines near Calais. Hazon literally describes 'the end of the hedge' from *heges ende* and referring to a boundary.

Hitchcroft probably represents 'the croft or smallholding of a man called Hicca', although the personal name in any example will almost always be uncertain to some degree. Rugley seems to come from 'the woodland clearing of a man called Rugga' and yet, with records of Rugeley in both 1267 and 1307, it could well share an origin with the Staffordshire town in *ruh leah* 'the rough woodland clearing'.

Coal mining began here in 1728 and, by the end of the century, six shafts had been sunk at what was previously Blue Lodge Farm but was known locally as Colliery Farm. When the pit was bought for the princely sum of £50 by the English Co-operative Wholesale Society at the beginning of the twentieth century, it resulted in a virtual new village of 170 homes, some intended for the retired miners, in an area still known as Colliers Close.

Shipley

Listed as Schepley in 1236, this is Old English *sceap leah* giving 'the woodland clearing where sheep are kept'.

Shotley Bridge

Two possible Old English origins for the basic name here. Either this is *sceote leah* and 'the woodland clearing frequented by pigeons' or *sceot leah* 'the woodland clearing where shooting takes place'. The name is recorded as Schotley in 1242, the addition is not found until some years later and is a reference to the vital crossing over the Derwent which forms the border with neighbouring County Durham.

Eddys Bridge seems to be speaking of 'the bridge of a woman called Edith', records show there was a woman of this name around here before the fifteenth century. Greymare Hill has nothing to do with horses, here the *grey maere* is a 'boundary stone'. Black Hedley describes the colour of the local soil in 'the dark woodland clearing of a man called Heathhere'. Hole Row identifies this as 'the row (of cottages) by the holes or hollows', while Orchardfield is self-explanatory. Allensford is recorded as Aleynforth in 1382 and

Allonsford in 1580, the name describing 'the ford of a man called Aleyn'. Allerside is found as Alarseth in 1261 and Allerseth in 1262, this is 'the slope growing with alder trees'.

Barley Hill comes from *bere hlaw*, Old English for 'the hill where barley is grown'. The modern addition of 'Hill' is thus superfluous. Birkenside is seen as Byrkinside in 1262 and as Brekenside in 1454, this being 'the hill growing with birch trees'. Cronkley is found as Crumclef in 1296 and Crumcliffe in 1298, this shows this is a corruption of Old English *crumbe clif* meaning 'the crooked cliff'. Crooked Oak has existed since at least the fourteenth century and is still a name of obvious meaning, this will have started out as a marker.

Shotton

A name recorded in the middle of the eleventh century as Scotadun, a name from Old English *Scot dun* telling us it was 'the hill of the Scots'.

Holford is a fairly common name and one which is either 'the ford in a hollow' or could warn of 'the hollows in the ford'.

Simonburn

The earliest surviving record of this name comes from 1229 as Simundeburn. Here the Saxon personal name and Old English *burna* combine to give 'the stream of a man called Sigemund'.

Barneystead is name from 'the enclosed pasture of a man called Beornhard'. Carriteth is one of the few place names in England to come from Old Norman French *caritedh* and speaking of 'the land used for charitable purposes', most likely associated with the church. Cragshiel still means exactly what it says, 'the 'shiel or shed by the crag'. Dally Castle comes from *dal leah* 'the common woodland clearing'.

Donkleywood features three Old English elements, *dun clif wudu* describing 'the wood at the steep hill'. From a Saxon

Simonburn's Millennium Bench

Mungo's Church, Simonburn

personal name and Old English *inga ham* comes a name meaning 'the homestead of the family or followers of a man called Eofel', which appears on modern maps as Ealingham. Early records of Gofton all show the suffix is Old English *dene* and not *tun* as the modern form suggets, thus this is 'the valley of a man called Gof'. Found as Haluton in 1177, Halton is from *halh tun* or 'the farmstead at the corner of land'.

Lych gate at St Mungo's Church, Simonburn

Humshaugh is found as Hounshale in 1279 and as

Hounshalgh in 1307, this being 'the *halh* or nook of land of a man called Hun'. Huntland holds few surprises in describing 'the land of the hunter'. From *nune wic* comes Nunwick, where the suffix is used to describe 'the special place of the nuns', the religious order owned this land but they were not resident.

Slaggyford

Found as Slaggiford in the thirteenth century, this is a Middle English place name which *slaggi ford* warns travellers it was a 'muddy ford'.

Slaley

Listed as Slaveleia in 1166, here is 'the muddy woodland clearing' from Old English *slaef leah*.

Bulbeck Common takes its name from Old Scandinavian and tells us this was 'the stream of a man called Bolli'. However neither the man nor the stream were ever here, for this is a name transferred from a village near the estuary of the Seine, chosen because it was the home of the knight who was given these lands in Northumberland by Henry I. Colpitts is one of the few place names which actually refers to the mining of coal, most references are to charcoal.

With the earliest listings of Dekesfeud in 1255 and as Dukesfeld in 1296, here is Old English *feld* with the personal name Ducc today known as Dukesfield. While *feld* is clearly the early form of the modern 'field', the only other similarity was this being 'open land' used for farming, for the Saxon *feld* was cleared land where the brushwood and shrubs were at the perimeter created by the cleared vegetation alone.

The name of Embley comes from Old English to describe 'the woodland clearing by the elm trees'. March Burn and Marchingley share a common origin, the latter probably came first as 'the woodland clearing of the family or followers of a man called Merc' and 'the stream of Merc' would then be an example of back formation. Shiel Hall is recorded since the thirteenth century, the name clearly showing this was built at 'the summer pastures'.

The Travellers Rest is an offer to passers-by to drop in for refreshment. The Rose and Crown is a patriotic establishment, the rose representing the country and the crown its monarch.

Sleekburn (East & West)

A name where the additions are self-explanatory, both having a common origin. Found as Sliceburne in the eleventh century, it is not clear if this Old English place name is *slicu burna* 'the smooth stream' or *slic burna* 'the muddy stream'.

At West Sleekburn we find the Foresters Arms, a reminder of the Ancient Order of Foresters, a well-established friendly society with meeting places (known as courts) in both the UK and USA.

Snitter

A highly descriptive name which is seen in a document from 1176 as Snitere. Here Middle English *sniteren* tells of the expected inclement weather at these latitudes is to be expected for this means 'to snow'.

Spindlestone

Seen as Spindlestan in 1187, this is 'the stone or rock said to resemble a spindle' and comes from Old English *spinele stan*.

Stamfordham

Despite the comparative complexity of this place name, the earliest record from 1188 is exactly the same as the modern form. Here Old English *stan ford ham* speaks of 'the homestead at the stone fort'.

Early records of Cheeseburn Grange does not feature the suffix *burna* but *burh*, thus is 'the manor known for its cheese'. The modern form of Kearsley would seem to suggest a suffix of *leah*, however records of Keerneslau in 1273, Kirneslawe in 1278, and as Kereslaw in 1361 show this was actually *hlaw* giving 'the hill of a man called Kjarni'. A name quite common over the border in Scotland, Mainsbank describes the *mains* or 'demesne lands on a bank of land'. Demesne is from French *demeine* and refers to lands not only under the control of the lord of the manor but used exclusively for his family's benefit. Bitchfield is recorded as Bechefeld in 1242 and as Bichfeld in 1421, this comes from Old English *bece feld* 'the open land where beech trees grow'.

Stannington

Found as Stanigton in 1242, this is from *staning dun*, Old English for 'the farmstead at the stony place'.

Blagdon took its name from *blaec dene* 'the dark valley', the water at its bottom trickled down in a small stream to join the Blyth. Clifton is among the most common minor place names, from *clif tun* this is 'the farmstead by a cliff or bank'. Milkhope is from Old English *meolc hop* 'the enclosed valley with rich pasture'. Plessey is probably a transferred name from France, Norman French *plessis* is the basis for over twenty place names and is derived from the Latin meaning 'to weave' and a reference to the appearance of countryside.

The Ridley Arms takes the name of the family who have resided at Blagdon Hall since 1720. The fountain in Church Road was donated by Viscount Ripley in 1902 to mark the coronation of Edward VII.

Stocksfield

From Old English *stoc feld* or 'the open land of the outlying place', this name is recorded as Stokesfeld in 1242.

The Duke of Wellington at Newton, the place name speaks for itself as 'the new *tun* or farmstead', remembers one of our greatest heroes. Arthur Wellesley, Duke of Wellington, is best known for his victory over Napoleon at Waterloo. However he later enjoyed a very successful political career, serving terms as both prime minister and home secretary.

Sunderland, North

Here the addition is to distinguish it from its larger and better known namesake to the south. Here the basic place name, listed as Sutherlannland in the twelfth century, is from Old English *sutherra land* and tells us it was 'the southern cultivated land'.

Swarland

Another place name where the earliest form is from 1242 and is recorded exactly as it appears today. Here Old English *swaer land* tells of 'the heavy cultivated land'.

Swinhoe

Recorded as Swinhou in 1242, this is Old English *swin hoh* and describes 'the headland frequented by swine'.

T

Thirston, East

Listed as Thrasfriston in 1242, here a Saxon personal name and Old English *tun* combine to talk of 'the farmstead of a man called Thraesfrith'.

Throphill

Found as Trophil in a document from 1196 and derived from Old English *throp hyll*, this place was once known as 'the hill of the outlying farmstead'.

Thropton

This name describes 'the farmstead with an outlying farmstead'. It is derived from Old English *throp tun* and is recorded as Tropton in 1177.

The local is the Cross Keys, an indication this probably stands on church land as it is named from the symbol of St Peter. However the local church is dedicated to St Andrew, so perhaps the sign already existed.

Titlington

Recorded as Titlingtona in the twelfth century, here is 'the farmstead associated with a man called Titel' and is derived from a Saxon personal name and Old English *ing tun*.

Togston

Listed as Toggesdena in 1130, the 'valley of a man called Tocg' sees a Saxon personal name followed by Old English *denu*.

Tosson, Great

With an addition to distinguish this from a similarly named minor place name, this is from Old English *tot stan* or 'the look-out stone' and is recorded as Tossan in 1205.

Tranwell

A name appearing as Trennewell in a document dated 1268. Telling of 'the stream frequented by cranes', this is a combination of Old Scandinavian *trani* and Old English *wella*.

Trewhitt, High

From a combination of Old Scandinavian *tyri* and Old English *wiht* comes 'the river bend where resinous wood is obtained'. First seen as simply Tirwit in 1160, the addition is to distinguish this from a minor place name and should be understood as 'more important' rather than any reference to elevation.

Tritlington

Here a Saxon personal name and Old English *ing tun* tell of 'the farmstead associated with a man called Tyrhtel'. The name is recorded as Turthlyngton towards the end of the twelfth century.

Hanging Leaves is a name which has nothing to do with trees. The first part tells us it is 'on a slope', the second is a plural of the word *laes* or 'pastures'.

Tweed (River)

This river name dates from before the Roman era, a Celtic or British name describing 'the powerful one'.

Tweedmouth

Today it is still easy to see this as 'the settlement at the mouth of the River Tweed'. The earliest record comes from 1208 as Tuedemue and features Old English *mutha* and a Celtic or pre-Celtic river name discussed in detail under its own entry. The local name of Ord is from *ord* meaning 'a point or corner of land'.

Pubs include the Harrow, an invitation to the agricultural works who would have recognised the image of an implement which is still used to break up the clods of earth created by the plough to a fine tilth. We must assume the slate roof of the Thatch House Tavern is an indication this is not how the original building looked.

Tyne, River

A Celtic river name meaning simply 'river', this is related to other major rivers in the country - Thames, Tean, Teme, Tene, Thame, Tamar - all with the same highly simplistic meaning.

River Tyne at Hexham

Tyne and Wear

A county formed by the upheavals of 1974. It is derived from the names of two rivers which loosely form the northern and southern boundaries. Two river names which mean 'river' and 'water' respectively.

Tynemouth

The earliest known record dates from 792 as Tinanmuthe and means exactly what it says, this is 'at the mouth of the River Tyne'. The river name is discussed under its own entry. The original Iron Age settlement was atop the headland overlooking the river's mouth when the place was known as Pen Bal Grag. Later occupied by the Romans, a monastery was established here in the seventh century. In 800 the Danes sacked the place which was promptly rebuilt only to suffer the same fate in 875, but was in use again by the time of the next record shortly after the Conquest.

Billy Mill is a place seen since the fourteenth century, although historically the name has referred to the two adjacent features of 'the hill of a man called Billy' and 'the mill of a man called Billy'. Chirton is a typical Saxon settlement name, featuring a personal name and the suffix *tun* this is 'the farmstead of a man called Ceorra'. From Old English *culfre cotes* comes Cullercoats, 'the dovecotes' which would have been a food source in Saxon times and later.

Flatworth's early records give the name as Flatford, describing 'the shallow ford', while the present form has been influenced by Old Scandinavian *vath* which means the same as Old English *ford.* From *maer denu* or 'the boundary valley' comes the modern name of Marden. There can be no arguments that Milton describes 'the farmstead with a mill'. Monkseaton describes 'the farmstead on the dry land in a marsh held by the monks'. Similarly Preston, from *preost tun*, speaks of 'the farmstead of the priests'.

The earliest record of Rodestane dates from 1320 and is exactly the same as today. Indeed to the Saxons this was little different, *rode stan* describing 'the rood stone', a marker for an important, probably religious, gathering or event. Kennes Dene has been recorded since the thirteenth century, this name describing 'the valley of a man called Cenwald'. Along the shoreline are rocks known as the Black Middens. Covered at high water these have seen many ships flounder here having considered themselves safely in the haven of the river entrance. An indication of the danger these proved can be seen in three days in 1864 when no less than five ships came to grief here, resulting in the loss of many lives. Their name ultimately comes from the term for a rubbish dump.

U

Ulgham

Listed as Wlacam in 1139 and as Ulweham in 1242, the latter record is closer to the original *ule hwamm* or 'the valley nook frequented by owls'. However that does not explain why this place, often referred to as "the village of the unpronounceable name", is known as 'Uffam'. It is still said to be 'the village of the owls', although mining has meant their habitat has been destroyed and these birds are rarely seen today. Yet the 'village of the owls' still holds true for nearly every home has an owl – be they indoors or out, stone, glass, wooden, china, bronze, or simply represented in paint or needlework.

Opposite the church, St John the Baptist a nineteenth century rebuild of the Saxon church, can be seen earth mounds which marked the original village boundary. Such must have been commonplace in Saxon times but these are thought to be the only surviving example in England.

W

Wall

A place found close to that most famous of Roman remains, Hadrian's Wall. Indeed it is this two thousand year-old structure which gives the place its name, via Old English *wall* and recorded as Wal in 1166.

It was here in 633 or 634 that Oswald of Bernicia led a victorious Northumbrian army against the Welsh under Cadwallon ap Cadfan of Gwynedd. Most often known as the Battle of Heavenfield, the Welsh recorded it as Bellum Cantscaul, and dated it as 631, while the Venerable Bede named it the Battle of Deniseburna near Hefenfelth. The name of the battle is not the only conflicting detail, reports differ wildly as to why it was fought, the numbers involved, and the strategies of each side.

Wansbeck, River

Early records of the name include Wenespic in 1139, Wanespik in 1255, Wanspic in 1271, Wanspyke in 1436, Wandesbeck in 1552, and Wanspek in 1610. The river rises at what is known as Wanny's Crags, possibly suggesting this is a name describing Wanny's Crags Water. There is also a Scandinavian stream name of *Vanspeck*, itself describing 'the stream of water' although there is no reason to suspect any connection between the two places.

Warden

Documented as Waredun towards the end of the twelfth century, here Old English *weard dun* refers to 'the look-out hill'.

Allerwash is from *alor waesc*, telling us it was 'the alluvial land marked by alder trees'. Such land saw regular seasonal flooding, bringing silt down from upstream which was deposited where the river burst its banks. This natural, if somewhat unpredictable, event brought nutrients to the land making it an excellent growing medium.

First seen in 1278 as Forstanes, Fourstones means exactly what it says 'four stones' which would have marked the boundary. Grindon is from *grenan dun* and tells of 'the green hill'. Haydon is found as Heiden in 1255, this being 'the valley where hay is cut'. Homer's Lane runs past that known for the 'marsh in a hollow'. Owmers is from Old English and describes 'the marsh frequented by owls'.

The Boatside Inn advertises its location.

Warenford

Given as Warneford in a document dated 1256, this tells us it was 'the ford over the Warren Burn'. Here a Celtic river name meaning 'the river where alder trees grow' is later joined by Old English *burna* or 'stream' and, later still, Old English *ford* or 'river crossing'.

Wark

One of two places of this name in the county. The example near Bellingham is listed in 1279 as Werke and, derived from Old English

Confluence of the local burn with the North Tyne at Wark

weorc, refers to 'the fortification'. It is often said to be Wark-on-Tweed.

Featuring a Saxon personal name and Old English *inga tun*, Hetherington speaks of 'the farmstead of the family or followers of a man called Haethhere'. Linacres is from *lin aecres*, literally 'flax acres' and telling us this was flax was grown, the word *lin* being the basis for 'linen'. Middleburn explains itself as 'the middle stream'. Shitlington is from 'the farmstead of the family or followers of a man called Scyttel', the Saxon personal name followed by Old English *inga tun*.

Wark

A second place of this name, this example is near Cornhill and found as Werch in 1158. As with its namesake above, this is Old English *weorc* or 'the fortification'.

The Grey Bull Inn was probably named to differentiate from the black version at Warkworth.

Warkworth

Recorded as Werceworthe in the middle of the eleventh century, here a Saxon personal name and Old English *worth* refer to 'the enclosure of a man called Weoca'.

Cheveley similarly unites a Saxon personal name and an Old English element. This time *leah* is the suffix for 'the woodland clearing of a man called Ceofa'. Gloster Hill is probably a Romano-Celtic name and, if so, then is identical to the early forms of Gloucester and speaks of 'the Roman town called Glevum', itself a Celtic name meaning 'the bright place'. Hadston tells of 'the *tun* or farmstead of a man called Hadd', a pet form of several personal names such as Heathustan.

In the case of Hauxley we know the meaning as 'the hawk's hill', however there are two quite different meanings for both elements and this could be 'burial mound' in which case the first element would be used as a personal name. Morwick, from *mor wic*, is recorded as Morewic in 1171 and as Morwick as early as 1278. The *wic* is normally defined as 'specialised farmstead' and most often this refers to a dairy farm. However here this simply does not seem to fit with the terrain, so perhaps this should be seen as 'dwelling by the swamp'. Sturton is derived from Old English *straet tun* 'the farmstead near the Roman road'.

Pubs named the Black Bull are undoubtedly taken from the powerful image adopted by many families, making it difficult to know the actual origin although the idea of it being religious and a reference to the papal bull cannot be dismissed. The church of St Lawrence is still mostly Norman, unique in the county, and was the scene of a massacre in 1174. Duncan, Earl of Fife, led a Scottish raid on the area and butchered some three hundred individuals.

Warton

Found in a document from 1236 as Wartun, this represents Old English *weard tun* or 'the watch out farmstead'.

Washington

Documented as Wessinton in 1197, Wassinton in 1211, Quessington in 1280, Wessington in 1311, and Weshington in 1507. This is probably a Saxon name from *inga tun* and telling of 'the farmstead of the family or followers of a man called Wassa'.

Locally we find Barmston, the 'farmstead of a man called Beorn'.

Westnewton

A name first seen in the twelfth century when it was given as Niewtona. The name is from Old English *niwe tun* or 'the newer farmstead' and with a self-explanatory addition to distinguish this from Kirknewton.

Whalton

Listed as Walton in 1203, here Old English *hwael tun* tells us it was known for being 'the farmstead by a round hill'. The Bale Fire, which features morris dancers and performances by children of the village outside their school, is an annual event held on Midsummer's Eve. That this is said to be 4th July is down to it dating from before the adoption of the Gregorian calendar in 1752, thus aligned with the earlier Julian calendar.

Locally the name of Riplington has been recorded since the thirteenth century. Here a Saxon personal name and Old English *inga tun* speaks of 'the farmstead of the family or followers of a man called Rippel'.

Whitfield

From Old English *hwit feld* and listed as Witefeld in 1254, this can still be seen as 'the white open land'. Just what was being described as 'white' is unclear, although it may be we should understand this as 'bright'.

The only thing certain about the name of Dingbell Hill is that it was corrupted earlier than the oldest surviving record of Vingvell hill in 1386. The most likely origin is seen for Thingwall in Cheshire, this meaning 'the field where an assembly meets' from Old Scandinavian *thing vollr*. Kingswood is one of the easiest names to understand, 'the woodland associated with the king'. Ulston takes its name from a Scandinavian personal name and Old English *tun* or 'the farmstead of a man called Ulkill'. With the earliest record as Parmontle in 1135 and Permanley in 1279, Parmentley describes 'the woodland clearing where pearmain grows', this being a variety of pear'.

Whittingham

Telling of 'the homestead of the family or followers of a man called Hwita' and from Old English *inga ham* following a Saxon personal name, this place is recorded as Hwittincham in the middle of the eleventh century.

The hamlet of Barton is from *bere tun* 'the farmstead where barley is grown'. Eslington is a Saxon personal name and Old English *inga tun* or 'the farmstead of the family or followers of a man called Aesc'. Lorbottle adds the suffix *botl* to a Saxon personal name to tell of 'the building associated with a man called Leofhere'. Place names nearly always describe the place and while 'building' might not seem particularly remarkable, we can be sure the building itself would have been much larger than most. Shawdon was recorded as Schaheden in 1232, as Schauden in 1250, as Shaweden in 1428, and Shawdon for the first time in 1542. This comes from Old English *sceaga denu* 'the valley with a small wood'.

Locals enjoy a glass of their favourite tipple at the Bridge of Aln. Named after the nearby crossing of the River Alne, a name discussed under its own entry.

Whittington, Great

Here a Saxon personal name and Old English *ing tun* combine to tell us it was known as 'the farmstead associated with a man called Hwita'.

The name is recorded as Witynton in 1233, while the addition, to distinguish this from a minor place name nearby, is not seen until much later.

Whittonstall

For the author to find a name which describes the place in such a way as to bring to mind an image of the place during Saxon times, an image which would never appear on any canvas. Here Old English *cwic tun stall* describes 'the farmstead bordered by a quickset hedge', the name recorded in a document of 1242 as Quictunsal.

Widdrington

Records of this name include Vuderintuna in the middle of the twelfth century, here is a Saxon personal name and Old English *ing tun* telling of 'the farmstead associated with a man called Widuhere'. The local pub takes the place name too.

Records of Chibburn are comparatively late for a name which was certainly coined in the Saxon era, for this is describing 'the stream of a man called Cilla'.

Wingates

Listed as Wyndegates in 1208, here Old English *wind geat* tells us of 'the windy pass'.

Maggleburn takes the name of the *burna* or stream, which already had a name of Celtic origins before the addition of the Saxon element for they did not understand the Maggle – and even today the name remains a mystery.

Woodburn, East

Documented as simply Wodeburn in 1265, this name describes 'the stream in the wood' from Old English *wudu burna*. The addition, to distinguish this from a similarly named minor place name, is found much later.

Woodhorn

A simple enough name meaning 'the wooded horn of land' from Old English *wudu horn*. The earliest surviving record of this name is as Wudehorn in 1178.

The minor place name of Druridge is unusual for the name is from *dryge hrycg* 'the dry ridge of land', which we would only expect to find in the southern counties of England. How it came to surface in the north of country will never be known. Hirst is the common suffix *hyrst* but is, as here, found alone describing 'the wooded hill'. Linton is a common name which can have a number of meanings, here this is 'the *tun* or farmstead on the River Lyne'.

Wooler

Seen as Wulloure in 1187, the 'spring promontory' is derived from Old English *wudu horn*.

Fenton is from Old English *fen tun* 'the farmstead by the fenland'.

Yeavering Bell is a nearby hill distinguished by its twin peaks. This was known as Din Gefron by the Brythonic peoples who lived in this hill fort, a tribe known as the Votadini, and a name related to Old Welsh speaking of 'the hillside of the goats'. The modern name is simply an Anglicised form of the earlier name. This site of twelve acres was enclosed by a stone wall ten feet thick with four clear entrances, each defended by a guard house. Votadini territory extended between the River Tyne and the Firth of Forth north of the border. This kingdom was known as *Gododdin* to the Welsh, this seen in the poem *Y Gododdin*, one of the oldest pieces of British literature known.

The Red Lion Inn has the most popular pub name in the land. As with all oddly coloured animal names this is heraldic but with the tremendous number of examples, thought to have peaked at something over 750, must refer to something or someone very important. If this is a something then it represents Scotland, which inherited the symbol from someone, that being John of Gaunt, the most powerful man in the country in the fourteenth century. Eventually his descendants of the House of Lancaster, including Henry IV, Henry V, and Henry VI, ruled England. Later the ruling houses of York and Tudor were also descendants of John of Gaunt. However John also had a number of illegitimate children as a younger man, the Beaufort line became rulers in Scotland and eventually, through James I, kings of England.

The Angel Inn has a more direct route, coming directly from the local church. While association between church and pub may seem odd today, when virtually everyone earned their living from the land these were the most likely places to meet your fellow villagers. Furthermore the pub often stood on land owned by the church, the largest landholder in the country after the monarch.

A less obvious religious reference is seen in the Anchor. While a maritime link is not impossible, the coast is just ten miles away, the vast majority were inspired by the words of St Paul in the book of Hebrews which notes how one's faith would be an anchor through life's troubled waters.

Wylam

Recorded as Wylum in the twelfth century, this is from Old English *wilum* and tells of the '(place of) the fish traps'.

No animal gave its name to the Black Bull Inn, this is undoubtedly taken from a relevant coat of arms. However this strong image has been chosen by so many it is impossible to know who is being commemorated. Yet in the case of the Fox and Hounds animals are certainly the basis, perhaps this was a meeting place for the local hunt. Location is behind the naming of the Boathouse Inn and also the Ship Inn, both found near the Tyne.

Y

Yetlington

Recorded as Yettlinton in 1187, this is derived from a Saxon personal name and Old English *ing tun* and tells of ‘the farmstead associated with a man called Geatela’.

Common place-name elements

Element	Origin	Meaning
ac	Old English	oak tree
banke	Old Scandinavian	bank, hill slope
bearu	Old English	grove, wood
bekkr	Old Scandinavian	stream
berg	Old Scandinavian	hill
birce	Old English	birch tree
brad	Old English	broad
broc	Old English	brook, stream
brycg	Old English	bridge
burh	Old English	fortified place
burna	Old English	stream
by	Old Scandinavian	farmstead
ceap	Old English	market
ceaster	Old English	Roman stronghold
cirice	Old English	church
clif	Old English	cliff, slope
cocc	Old English	woodcock
cot	Old English	cottage
cumb	Old English	valley
cweorn	Old English	queorn
cyning	Old English	king
dael	Old English	valley
dalr	Old Scandinavian	valley
denu	Old English	valley
draeg	Old English	portage
dun	Old English	hill
ea	Old English	river
east	Old English	east
ecg	Old English	edge
eg	Old English	island
eorl	Old English	nobleman
eowestre	Old English	fold for sheep

fald	Old English	animal enclosure
feld	Old English	open land
ford	Old English	river crossing
ful	Old English	foul, dirty
geard	Old English	yard
geat	Old English	gap, pass
haeg	Old English	enclosure
haeth	Old English	heath
haga	Old English	hedged enclosure
halh	Old English	nook of land
ham	Old English	homestead
hamm	Old English	river meadow
heah	Old English	high, chief
hlaw	Old English	tumulus, mound
hoh	Old English	hill spur
hop	Old English	enclosed valley
hrycg	Old English	ridge
hwaete	Old English	wheat
hwit	Old English	white
hyll	Old English	hill
lacu	Old English	stream, water course
lang	Old English	long
langr	Old Scandinavian	long
leah	Old English	woodland clearing
lytel	Old English	little
meos	Old English	moss
mere	Old English	lake
middel	Old English	middle
mor	Old English	moorland
myln	Old English	mill
niwe	Old English	new
north	Old English	north
ofer	Old English	bank, ridge
pol	Old English	pool, pond
preost	Old English	priest
ruh	Old English	rough
salh	Old English	willow
sceaga	Old English	small wood, copse
sceap	Old English	sheep
stan	Old English	stone, boundary stone

steinn	Old Scandinavian	stone, boundary stone
stapol	Old English	post, pillar
stoc	Old English	secondary or special settlement
stocc	Old English	stump, log
stow	Old English	assembly or holy place
straet	Old English	Roman road
suth	Old English	south
thorp	Old Scandinavian	outlying farmstead
treow	Old English	tree, post
tun	Old English	farmstead
wald	Old English	woodland, forest
wella	Old English	spring, stream
west	Old English	west
wic	Old English	specialised, usually dairy farm
withig	Old English	willow tree
worth	Old English	an enclosure
wudu	Old English	wood

Bibliography

Dunkling, Leslie and Wright, Gordon *A Dictionary of Pub Names*
Ekwall, Eilert *The Concise Oxford Dictionary of English Place-Names*
Leslie, Jack and John *Down Our Streets*
Mills, A.D. *Oxford Dictionary of English Place Names*
Newcastle upon Tyne City Libraries and Arts *What's In A Name?*

Also from Sigma Leisure:

County Durham Place Names
Anthony Poulton-Smith

Ever wondered why our towns and villages are named as they are? Who named them and why? Which place seems to refer to dairy farms but actually refers to fishermen? Was Crook founded by a criminal? And how does anywhere get a name like Pity Me?

Within these pages author Anthony Poulton-Smith examines the origins of the place names with which we are otherwise so familiar. Towns, villages, districts, hills, streams, woods, farms, fields, streets and even pubs are examined and explained. Some of the definitions give a glimpse of life in the earlier days of the settlement, and for the author there is nothing more satisfying than finding a name which gives such a snapshot. The definitions are supported by anecdotal evidence, bring to life the individuals and events which have influenced the places and how these names have developed.

This is not simply a dictionary but a history and will prove invaluable not only for those who live and work in the county but also visitors and tourists, historians and former inhabitants, indeed anyone with an interest in County Durham.

£8.99

Cambridgeshire Place Names
Anthony Poulton-Smith

Ever wondered why our towns and villages are named as they are? Who named them and why? Within these pages author Anthony Poulton-Smith examines the origins of the place names with which we are otherwise so familiar. Towns, villages, districts, hills, streams, woods, farms, fields, streets and even pubs are examined and explained. Some of the definitions give a glimpse of life in the earlier days of the settlement, and for the author there is nothing more satisfying than finding a name which gives such a snapshot. The definitions are supported by anecdotal evidence, bring to life the individuals and events which have influenced the places and the way these names have developed.

This is not simply a dictionary but a history and will prove invaluable not only for those who live and work in the county but also visitors and tourists, historians and former inhabitants, indeed anyone with an interest in Cambridgeshire.

£9.99

You Take the High Road
Northumberland Walks
With alternative routes to suit all abilities
Anne Leuchars & Debby Waldron

This book proves that the tortoise and the hare can set off and finish at the same time, and enjoy each other's company along the way. Anne Leuchars is the tortoise, a slow walker who likes to stop for a breather, and Debby Waldron is the hare, fit enough for any outdoor challenge. Together they have devised walks which bring equal pleasure to both types of walker. Each walk splits into harder sections for fitter people and easier tracks for slower walkers. In most cases you all set off together, then separate part way through and meet up again to complete the walk together. The walks also accommodate differences in approach to a day in the countryside. Some people like the satisfaction of several miles or a steep hill conquered, others prefer to stroll along, taking rests along the way to relish the views.

8.99

Discovering NewcastleGateshead
A walking guide to NewcastleGateshead plus parkland strolls and visits to coastal attractions
Peter Donaghy & John Laidler

26 carefully crafted walks lead to places of historical, architectural and social interest, at times through nooks and crannies that may be new even to long-standing residents. The walks are of varying lengths with opportunities for internal visits and refreshments. To involve children in the discovery process, a number of quiz questions have been included.

£10.99

Walking the Cheviots
Classic Circular Routes
Edward Baker

This book provides an excellent introduction to this solitary, wild countryside. Everyone is catered for from weekend family walkers to the experienced hill walker with all the walks personally checked and trod by the author. Each route contains details of the natural history, geology and archaeology of the area descriptively written by Edward Baker who has lived his whole life in the region. For ease of reference, the book is in two sections, covering the northern and southern Cheviots - distinct areas with their own unique character.

There are almost 50 walks - by far the most comprehensive collection published for the Cheviots.

£9.99

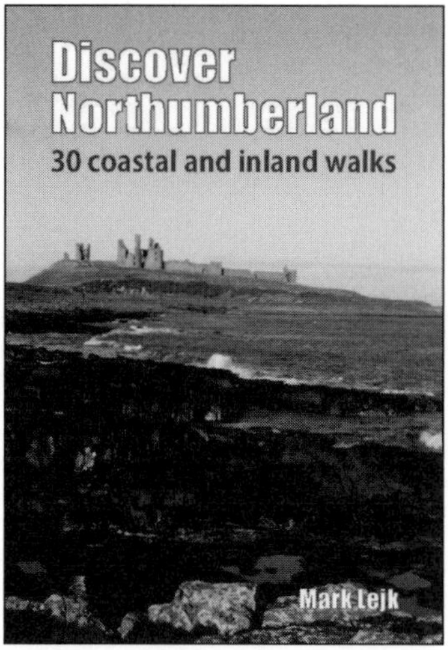

Discover Northumberland

30 coastal and inland walks

Mark Lejk

30 circular walks covering the whole of Northumberland from north to south and east to west aimed at people of all walking abilities and preferences with lengths ranging from 4.5 km (2.8 miles) to 17.8 km (11.1 miles) and difficulty ranging from easy to strenuous. Many of the longer walks have shorter versions and altogether the different variations and combinations make for 48 routes. The walks range from the fabulous Northumberland coastline to the Cheviot Hills, Kielder Forest, the North Pennines, Hadrian's Wall and there are a significant number of walks in the area between the coast and the Cheviots.

£8.99

Walking the Northumberland Dales

Hadrian's Wall Country

Norderhaug & Thompson

This guidebook explores the lesser-known landscapes of North and South Tynedale, Allendale, Hexhamshire, Blanchland and Hadrian's Wall, all within easy access of Newcastle-upon-Tyne, Durham and other popular locations. The walks are packed with interest on the history, industrial archaeology, traditions, and culture of the area.

£7.95

Adventure Walks for Little Explorers

in the Yorkshire Dales

Rebecca Chippindale and Rebecca Terry

This isn't just a book of fifteen walks for small children – it's an all-inclusive family adventure. Grab your 'Kit' and prepare to open up your senses. Feel parts of the journey, magnify the Yorkshire Dales, collect items along the way and create something different when you get home.

£8.99

All-Terrain Pushchair Walks North York Moors

Rebecca Terry and Zoë Sayer

The latest in the increasingly popular series of All-Terrain Pushchair Walks books, experienced mums, authors and walkers Rebecca and Zoë offer a selection of thirty all-terrain pushchair walks in the North York Moors National Park incorporating wild moorland, wooded valleys and spectacular coasts. The authors' at-a-glance symbols giving accurate grading for each of the walks making it easy to select the best one for you.

£8.99

North Yorkshire Coast Holiday Walks

Brian Conduit

The 20 walks range from 2½ to 6½ miles, a well as varying in length, they vary in difficulty and the nature of the terrain and some are inevitably more strenuous and involve more hill climbing than others.

£8.99

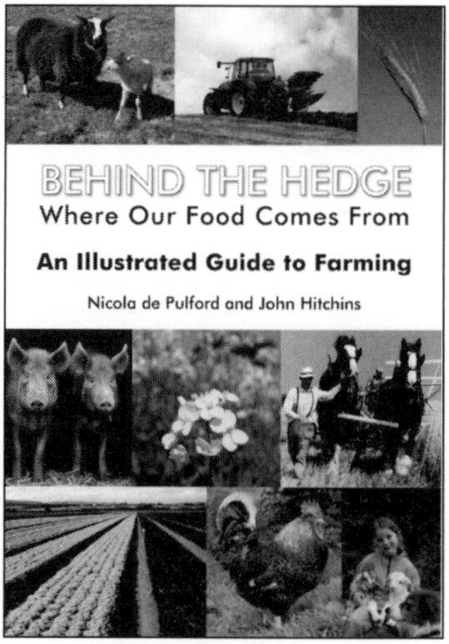

Behind the Hedge

Where Our Food Comes From An Illustrated Guide to Farming

Nicola de Pulford & John Hitchins

Behind the Hedge is for everyone who wants to know more about the food we eat, the land it is grown and reared on, and those who farm it. It is an easy-to-follow guide which will help you identify in their natural environment our crops, fruit and farm animals, agricultural buildings and machinery, the farming landscape and the wildlife it supports.

Never again will you mistake a field of wheat for one of barley, or an Aberdeen Angus cow for a Hereford. By dipping in and out of this beautifully illustrated book, you will learn to recognise the crops, farm animals and wildlife on the other side of the hedge.

£12.99

All of our books are available through booksellers.
For a free catalogue, please contact:

Sigma Leisure, Stobart House, Pontyclerc
Penybanc Road, Ammanford SA18 3HP

Tel: 01269 593100 Fax: 01269 596116

info@sigmapress.co.uk **www.sigmapress.co.uk**